Bitcoin, Blockchain & Fintech

Dr. Len Mei

What is blockchain?

Bitcoin, Blockchain & Fintech

Preface

Bitcoin, the first cryptocurrency, is probably the most explosive asset in the human history. In its short nine-year history, from 2009 to 2017, its market capitalization has grown from zero to almost US$200 billion at the end of 2017.[1] Since Bitcoin appeared, there are many other cryptocurrencies springing to life. At current count, there are over 1,400 different cryptocurrencies. The top five cryptocurrencies have combined market capitalization of over US$300 billion at one time.

Many people consider it is nothing more than a bubble. Indeed, the value of cryptocurrencies has endured violent ups-and-downs. However, such an explosive growth is more than just speculation. Its underlying technology, the blockchain, shows enormous potentials. Regardless the future value of cryptocurrency, the blockchain technology is here to stay. We are at the beginning of a digital revolution. Blockchain together with the other nascent technologies, such as artificial intelligence, Big Data, IoT and many others are the driving force of this coming digital revolution.

Blockchain technology comes from cryptography – a branch of computer science. What makes blockchain technology unique from the traditional cryptography is that it merged with the internet technology and transformed the internet from a system of information and data transfer and storage to become a system of asset transfer and storage. Suddenly, internet finds itself capable of performing many financial functions with the security and speed unmatched by the traditional financial systems. Bank of England

---

[1] https://coinmarketcap.com/

What is blockchain?

released a report on how to use the blockchain technology to strengthen its real-time settlement system, Swift proposed block-linked roadmap; IMF released the first digital currency report. Gartner estimates that blockchain could produce US$176 billion in business value by 2025.[2]

The birth of blockchain technology in the first decade of the 21st century is not by accident. It is due to the maturity of internet technology itself, the advance in computer science, the spread of cheap computing power, the high speed, high bandwidth communication and many other factors such as the e-commerce, and trade globalization.

The blockchain technology is still at a nascent stage. It is evolving and changing. Hundreds of startups and large companies are experimenting with it for the development of new applications. Its application in the cryptocurrency caught the most attention of the news media; however, its other applications will have a much more profound impact eventually.

Blockchain technology is revolutionary because, for the first time, we can transfer the objects with a monetary value over the internet and store them in the database securely and quickly. Before the invention of Bitcoin in 2009, nobody could figure out a way to make a digital currency work. Since the replication of data on the network is easy, such data cannot represent an asset. This is the double-spending problem. This changed when a person called Satoshi Nakamoto invented the Bitcoin using blockchain technology in 2009. If the blockchain technology can create and transact digital currency, it can also create and transact digital assets. The blockchain-based smart contracts and distributed apps open up a wide frontier for transaction applications. In the same way, the blockchain can also create an immutable digital identity, which provides chain of custody, proof of the ownership of assets. All the elements necessary to conduct all sorts of financial activities online are in place.

However, the blockchain technology is not the only technology that is altering many aspects of the financial, business,

---

[2] "Predicts 2018: top predictions in blockchain business", https://www.gartner.com/doc/3827065/predicts--top-predictions-blockchain

trade systems today. It is the combinational deployment of the artificial intelligence, blockchain technology, Big Data, communication, internet technology, computing technology and many others to cause a drastic shift in the way to do things.

Simultaneously, with the development of blockchain technology, there is also a new financial technology called Fintech. Fintech promises to revolutionize the financial industry just like automation to revolutionize manufacturing. These new technologies can add trillions of dollars to the global economy. The implications are staggering.

This book discusses some basic concepts and elements of the blockchain technology, their issues in development and potential applications. We will discuss the financial technology or Fintech, its status, and applications. This subject is so vast and quickly changing. It is impossible to cover the subject entirely in one book. This book serves as a comprehensive introduction and background to who is interested in the subject to do further research on the subject. As much references are cited as possible for the ease of research.

April 2018 California

Bitcoin, Blockchain & Fintech

Table of Contents

Preface 3

Chapter 1 What is blockchain? 13

    Section 1.01   Introduction 13

    Section 1.02   Toward the distributed computing 17

    Section 1.03   Centralization vs. Decentralization 18

    Section 1.04   Bitcoin 25

    Section 1.05   The proliferation of blockchain technology 29

    Section 1.06   Initial Coin Offering (ICO) 30

    Section 1.07   Blockchain platforms 32

Chapter 2 Bitcoin 35

    Section 2.01   Bitcoin, a ledger 37

    Section 2.02   How does Bitcoin work? 39

    Section 2.03   Digital signature 41

    Section 2.04   Hash 44

    Section 2.05   Merkle tree and block header 48

    Section 2.06   Difficulty and nonce 50

    Section 2.07   Bitcoin supply 54

    Section 2.08   P2PKH and P2SH Bitcoin addresses 55

    Section 2.09   Zero Knowledge Proof (ZKP) 58

    Section 2.10   Divisible Bitcoins 59

    Section 2.11   Keep your digital assets safe 61

    Section 2.12   Two Factor Authentication (2FA) 67

Chapter 3 Bitcoin issues 71

What is blockchain?

Section 3.01	Bitcoin issues	72
Section 3.02	Bitcoin and decentralization	76
Section 3.03	MtGox incident	78
Section 3.04	Full nodes vs. partial nodes	79
Section 3.05	Is Bitcoin truly anonymous?	81
Section 3.06	Transaction fee	82
Section 3.07	Transaction malleability	84
Section 3.08	BIP, Hardfork, and Softfork	86
Section 3.09	UASF and UAHF	88
Section 3.10	SegWit and change of block size	91
Section 3.11	Bitcoin split	93
Section 3.12	Keep your coins safe during forking	96
Section 3.13	Other proposed fixes	98
Section 3.14	Lightning network	99
Section 3.15	Government attitude	103

Chapter 4   Consensus mechanisms 107

Section 4.01	Proof of Work vs. Proof of Stake	109
Section 4.02	PoW and PoS hybrid	112
Section 4.03	dBFT – an alternative to PoW and PoS	114
Section 4.04	Paxos and Raft	117
Section 4.05	Proof of Concept	120

Chapter 5   Altcoins 121

Section 5.01	Litecoins – a lighter version of Bitcoin	122
Section 5.02	Zcash – a token with privacy	123
Section 5.03	Ripple – a digital equivalent of SWIFT	124

Section 5.04	Ethereum: the smart contract blockchain 131	
Section 5.05	DAO hack and Ethereum fork	138
Section 5.06	Legal issues	141
Section 5.07	DApps – Decentralized Apps	142
Section 5.08	Create your own coins out of Bitcoin	144
Section 5.09	Antshares or NEO – a multi-use token	146
Section 5.10	Bridging Bitcoin & EVM	149
Section 5.11	Asset digitization	150

Chapter 6   Other blockchain platforms.......................155

Section 6.01	Permissioned vs. Permissionless	155
Section 6.02	Identity, Transaction and Content MDL's 157	
Section 6.03	Identity MDL	161
Section 6.04	Digital currencies	166
Section 6.05	Token-less MDLs	170
Section 6.06	BigchainDB	171
Section 6.07	Corda	173
Section 6.08	HydraChain	176
Section 6.09	MultiChain	176
Section 6.10	Quorum	179
Section 6.11	Hyperledger	180
Section 6.12	Decentralized internet	183
Section 6.13	Other blockchain platforms	185
Section 6.14	Beyond blockchain	187

Chapter 7   Financial blockchain applications 189

Section 7.01	Banking and payment	189

What is blockchain?

Section 7.02	Credit card and loan applications	191
Section 7.03	Insurance	193
Section 7.04	Security trading	195
Section 7.05	Commodity trading	198
Section 7.06	Energy trading	200
Section 7.07	The sharing economy	202
Section 7.08	Peer-to-peer lending	203
Section 7.09	Cross-border retail business	205
Section 7.10	Counterparty platform DApps	207
Section 7.11	AML & KYC	209

Chapter 8   Other blockchain applications 213

Section 8.01	Data management	213
Section 8.02	Distributed data storage & ERP service	215
Section 8.03	Forecasting and prediction	217
Section 8.04	Internet of Things	219
Section 8.05	IOTA and IoT Chain	223
Section 8.06	Supply chain management	226
Section 8.07	Governance and voting	229
Section 8.08	Regulatory applications	233
Section 8.09	Land title registration and real estate	235
Section 8.10	Law and blockchain	237
Section 8.11	Protection of intellectual property	238
Section 8.12	Healthcare	240
Section 8.13	Food safety	242
Section 8.14	Defense industry	243
Section 8.15	Golem – a world computer	245

Chapter 9 Fintech & the future of financial market ... 250

Section 9.01   Financial vs. technology companies 254

Section 9.02   Global Fintech landscape 256

Section 9.03   Technology-driven Fintech 258

Section 9.04   Build MDL for financial services 264

Section 9.05   Blockchain technology for banks.......... 267

Section 9.06   Mobile wallet 269

Section 9.07   Crowdsales 270

Section 9.08   Wealth management 271

Section 9.09   A European example - T2S.................... 273

Section 9.10   TechFin 275

Section 9.11   Alternative Trading System 277

Section 9.12   O2O business 279

Section 9.13   Fintech and AI.. 282

Chapter 10   Fintech in China 292

Section 10.01   Major Fintech companies in China 298

Section 10.02   Chinese Fintech landscape 303

Section 10.03   Chinese credit rating system 308

Section 10.04   Chinese online lending market........... 314

Section 10.05   Chinese wealth management, e-insurance, and online trading 318

Section 10.06   Chinese third-party payment market. 325

Section 10.07   Chinese healthcare market 334

Section 10.08   China digital currency policy 334

Section 10.09   Regulating TPP in China 336

Chapter 11   A glimpse of the future 339

What is blockchain?

Section 11.01	Pillars of the future technology evolution 341	
Section 11.02	Data drive Fintech	343
Section 11.03	Communication driven Fintech	346
Section 11.04	Conclusion	348

Figures **351**

Index **353**

Chapter 1   What is blockchain?

Section 1.01   Introduction

Blockchain is a technology to transact, distribute and store digital information on the internet. It does so by linking blocks of data and distribute them across a network. The data in blockchains are secure and immutable. Its first application is to create a digital currency – Bitcoin.

However, the technology is now finding many other potential uses, as specific as product traceability, copyright protection, any kind of financial transactions, and as broad as entertainment, publishing, energy, healthcare and many other industries.[3] It can be used across the entire value chain, benefiting businesses and consumers alike. Blockchain has the potential to transform the way that individuals and organizations interact, the way that businesses collaborate with one another, the transparency of processes and data, and ultimately, the productivity and sustainability of our economy. With its capability of guaranteeing data security, blockchain technology will be a foundation for any industry in which data security and trust are must.

A typical blockchain application involves several basic components: participants, assets, access control and transactions. Its value chain ranges from hardware manufacturing, platform and

---

[3] "What is a blockchain", Nick Williamson, https://www.finyear.com/What-is-a-Blockchain_a34985.html

What is blockchain?

security to application services, investment, media and human resources.

Since blockchain resides on the internet and its objective is to move "valuable assets", whether it is money or contract, it is also known as the Internet of Value (IOV). [4] The difference between the internet as we know today and the internet of value is the asset value of the contents. The internet today can move duplicated copies of data around. These data are for information only. The data in blockchain applications are either original or verified copy of asset that is as good as the original. When the blockchain technologies spread, more and more applications are on the horizon. Many of these applications target for the financial uses. These applications are known as Fintech, short for financial technology. We will discuss Fintech in more detail in the later chapters of this book.

In fact, many of the uses are revolutionary and create paradigm shifts. For example, blockchain technology can automate financial industries, much as robots automate manufacturing. Banks, stock exchanges, insurance companies, supply chains, land registration are therefore experimenting with blockchain technology. Other potential blockchain applications include medical records management, voting tracking, identity management and protection, long-term record storage, chain-of-custody for insurance policies and any types of legal documents such as real estate ownership, IoT usage tracking for industry and logistics, and for virtually every kind of digital record and transaction.

Another name for a blockchain technology is a Mutual Distributed Ledger (MDL) technology, or Distributed Ledger Technology (DLT). A ledger is a chronological record of transactions. The word "Mutual" denotes the shared nature of the distributed ledger. It is a special type of the more generalized distributed ledger.

In addition to being a distributed ledger, the blockchain is also a distributed database. It enables multiple parties to share and

---

[4] "The Internet of Value – Exchange", https://www2.deloitte.com/content/dam/Deloitte/uk/Documents/Innovation/deloitte-uk-internet-of-value-exchange.pdf

update data in a safe and secure way even if they do not trust each other. The blockchain fills in the missing piece of the internet today that is the transfer of assets online securely. It finds itself in wide varieties of applications to perform transactions or exchange of values safely and securely.

Some fundamental construction mechanisms of the blockchain, when first introduced in 2009, make the blockchain technology unique: First, it is distributed in a P2P network. It is also decentralized and encrypted.

The signed, encrypted, validated new transactions are bundled together into a "block". The new block links to the previous verified block in the blockchain cryptographically. In such a way, the blocks are forming a permanent and immutable chain of records. It is as if each new page of the ledger, when filled, is bound to the previous page of the ledger. The balance at the beginning of a new page must match that of the end of the previous page. In this way, the balances on each page form a chain with the balances of previous pages. Multiple copies of the identical ledger books will receive a duplicated new page so that all the ledger books will be identical after attaching the new page. In doing so, it keeps many identical copies of a record of the transaction history.

Identical copies of the blockchain are replicated across all nodes in the network. All copies of the blockchain automatically update themselves to be identical every 10 minutes. During this interval, different versions of blockchain may appear in different nodes. This phenomenon is called forking. In most cases, fork dissipates quickly and automatically.

The nodes communicate over a network and collaboratively construct the blockchain without relying on a central authority. Faults can appear when individual nodes crash, the communication between nodes fail, or there is a malicious attack. However, because there are many more nodes, the record remains safe. Such a safe record keeping is achieved by a consensus protocol. That is, the identical copies kept by majority nodes are considered as the authenticated copy.

The number of nodes is so vast that it is impossible to alter the transaction replicated in hundreds of thousands of sites in the

What is blockchain?

network. Such a regular update deprives any attempt to alter the data held across multiple nodes horizontally and integrated vertically along the chain. Any attempt will inevitably create a discrepancy or inconsistency among the data in the sequential blocks and different copies of the blockchain. This is how it becomes immutable.

The blockchain is also accessible to anyone on the internet. Therefore, it can be regarded as a public internet ledger.

All these copies of the blocks are more than just secure sequential distributed databases. Because they can be trusted, they offer the ability to simplify and automate transactions and make it easier and cheaper to conduct financial activities. As such, blockchain technology has gained a reputation for Trust and Transaction Machine.

The blockchain resides on the internet. In many ways, blockchain network is similar to the internet. On the internet, the web servers or HTTP servers serve web applications. In the same way, in the blockchain network, the so-called consensus servers serve distributed applications. The application sends a transaction to the platform, which handles communication and consensus. When the community agrees on the order of transactions, it then changes the state, which is recorded in the distributed database.

Bitcoin is the first application of blockchain technology. The nature of P2P distribution and decentralization is true to the Bitcoin blockchain and some initial blockchain applications. As Bitcoin starts to proliferate, interest in blockchain technology grows. In less than ten years since the idea of Bitcoin was born, countless industries from financial services to healthcare have begun contemplating how to leverage the technology for their own uses.

The distributed nature, validation and encryption methods and the trust mechanisms are all evolving to meet the purpose of the particular applications of the developers.

As more and more applications evolve, the basic mechanism of constructing blockchain also evolves. Some of its derivatives can be far from the idealistic distributed, decentralized, the trust machine originally invented for Bitcoin. However, it does not diminish the importance of blockchain technology in any way.

It contributes to creating a completely new technology called Fintech, short for financial technology, which would not be so salient without the blockchain technology.

Section 1.02   Toward the distributed computing

Modern computing infrastructure is becoming more and more distributed for many reasons. First, the price of CPU/ GPU and storage media (memories and disk drive) is becoming extremely cheap. They provide computing power and data storage to everyone, even the poorest. Today, a smartphone has more computing power than all NASA had in 1969.[5] The widespread computing power generates higher and higher volumes of data and forces us to handle data in parallel. The cheap storage also encourages data safeguarding by replication. When one of the storages fails, the valuable data are not lost. Furthermore, the ever-increasing communication speed and bandwidth make remote services not only possible but also necessary. Almost all the online applications access data remotely and process these data either locally or remotely. In this way, the distributed computing power co-exists with the distributed database.[6]

When multiple computing powers are processing the same dataset, they may generate different outputs. For example, in searching the same keywords in Google on different days, it may give different results. In some applications, this may not be acceptable. To avoid the conflict, the distributed systems need a consensus mechanism. If not, the distributed systems will soon enter the chaotic situation. In another word, consensus mechanism is a fundamental building block in the distributed systems. It guarantees the consistency of the outcome of whatever process in the system. It is the collective agreement on some value. Consensus makes it possible that a distributed system acts as a

---

[5] http://www.northeastern.edu/levelblog/2016/04/21/smartphones-supercomputer-in-your-pocket/
[6] "Distributed Computing: Fundamentals, Simulations and Advanced Topics", H. Attiya et.al. , Wiley, second edition, 2004.

What is blockchain?

single entity. A distributed computing system without consensus is like a city without traffic lights. The traffic will be chaotic.

Section 1.03 Centralization vs. Decentralization

The words of decentralization and distribution do not have a clear definition in the context of blockchain technology. As the technology evolves, the blockchain technology also evolves into different directions, which fall into the undefined territory of decentralization and centralization. Vitalik Buterin, the founder of Ethereum, another popular cryptocurrency, defined[7] centralization through three concepts: architectural, political and logical. Architectural centralization refers to the database distribution nature of the ledger. Political centralization (or simply centralization) refers to the ownership or presence of a central authority. While logical centralization refers to the state and the behavior of the blockchain. Vitalik Buterin considered blockchain politically and architecturally decentralized, but logically centralized. This is true for certain types of the blockchain.

When a blockchain is not truly public and has an owner, it is called the Permissioned blockchain.[8] The Permissioned blockchain is not politically decentralized because its owner controls it. To access a Permissioned blockchain for transaction, validation, and other privileges require permission from someone, usually the owner/ administrator of the blockchain. The owner/

---

[7] "The meaning of decentralization", Vitalik Buterin, Medium, https://medium.com/@VitalikButerin/the-meaning-of-decentralization-a0c92b76a274

[8] "Consensus-as-a-service: a brief report on the emergence of permissioned, distributed ledger systems", Tim Swanson, Great Wall of Numbers, http://www.ofnumbers.com/2015/04/06/consensus-as-a-service-a-brief-report-on-the-emergence-of-permissioned-distributed-ledger-systems/

Bitcoin, Blockchain & Fintech

administrator retain control over how it operates.[9] Since there are many privileges, which have the different level of centralization, this makes the exact boundary between decentralization and centralization fuzzy at best.

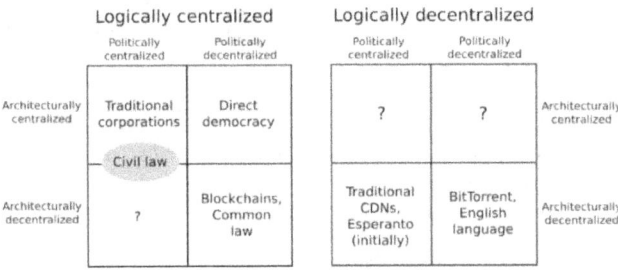

Figure 1-1 Definition of centralization by Vitalik Buterin

The Permissioned blockchain is also called the private blockchain, or consortium blockchain. This is because the designer/ creator of the Permissioned blockchain usually has some specific object or purpose in mind for its application. There are many Permissioned blockchain platforms, just to mention a few: Hyperledger Fabric; Tendermint; Symbiont Assembly; R3 Corda; Iroha; Kadena; Chain; Quorum; and MultiChain.

To differentiate the centralized nature of the Permissioned blockchain, the originally developed Bitcoin-like decentralized P2P blockchain is now called the Permissionless blockchain, or the public blockchain. The difference between the Permissionless or public blockchain and Permissioned or private blockchain is akin to that of the internet and the intranet. It can also be compared to the public park and private garden. The former belongs to the public, while the later has a private owner.

In a Permissionless blockchain, such as Bitcoin blockchain, anyone can be a user or run a node; anyone can invoke transactions, and participate in the consensus process. On the other

---

[9] "What has been the reaction to permissioned distributed ledgers?", http://www.ofnumbers.com/2015/04/26/what-has-been-the-reaction-to-permissioned-distributed-ledgers/

What is blockchain?

hand, in the Permissioned blockchain, the participation is by invitation only.

A Permissioned blockchain can be distributed but not necessarily decentralized. Even a centralized Permission blockchain still offers benefits over a traditional centralized database, although, the claim is sometimes rebuked by critics. Some of the benefits are:

- Each participant retains full control over assets;
- The data are encrypted and cannot be altered easily, therefore, safe;
- Alterations are traceable;
- Control of the database is distributed across entities;
- The loss of any one server would not compromise the network.

The table below lists the comparison of these three types of the database:

Property	Permissionless blockchain	Permissioned blockchain	Traditional database
(Political) Centralization	Decentralized	Centralized	Centralized
Distribution	Distributed	Distributed	Non-distributed or limitedly distributed
Validation	Public	Private	Private
Access	Public	Private	Private
Ownership	Public	Private	Private
Trust mechanism	Cryptology	Cryptology and/or 3rd party	3rd party

The word "Centralization" has a different meaning from the word "Distribution". A distributed database resides in many computer nodes, where a non-distributed database resides in one location. On the other hand, a centralized database has an owner or central authority overseeing the database, while a decentralized database does not.

There are different degrees of the centralization or decentralization. For example, in the Bitcoin or Ethereum platform, miners can create coins and anybody can be a miner. This is distinguishably decentralized vs. a national digital currency, which can only be created by the government. However, when compared to Bitcoin, Ethereum is more centralized.

The Bitcoin platform can create up to 21 million Bitcoins as coded in the software. Nobody can change the protocol unless there is a consensus in the Bitcoin community. This is decentralization. On the other hand, the maximum number of ETH is around 100 million, which has also been codified. However, Ethereum Foundation holds the power to change the protocol including the cap for its coins. In this aspect, Ethereum is less decentralized than Bitcoin.

Nevertheless, Ethereum is a platform, which allows anyone to develop an application on it. There are multiple clients written by independent parties. Ethereum Foundation has control over its own client only. Furthermore, if Ethereum Foundation forces to activate a change in its own client, and some people do not agree, they can split off as a hardfork. This is what happened in the Ethereum and Ethereum Classic split. From this point of view, Ethereum is still considered as decentralized.

At the other end of the spectrum of centralization, there is national digital currency. The government decides when to print money or how much. No dissident can split a hardfork of a national currency.

Figure 1-2 shows the degree of centralization for three criteria: authority, power, and control. There is no such thing as a 100% decentralized system or centralized system. Each system has its own degree of centralization for authority, power and control independently represented by lines AB, AC and AD. Point E indicates the degree of centralization of authority. Likewise, Point F is for power and point G is for control. When the point is closer to the apex, the degree of centralization increases. Therefore, we can use a triangle formed by the vertexes of three lines connecting

What is blockchain?

three points along the axis of power, control, and authorization (triangle EFG) to represent the degree of decentralization.

The Permissioned blockchain development is very important because most of the Fintech applications are this type of blockchain. Some decentralized properties for Permissionless blockchains do not apply to the Permissioned blockchain. The Permissioned blockchains do have owners, who act as centralized authorities to the blockchains. These two branches of blockchains are evolving in very different directions. They have a very different philosophy and applications. This will be clear in the latter discussion in this book.

Both types of blockchains guarantee the validity of transactions by recording it in a connected distributed system of registers with a secure validation mechanism. Most blockchain applications reside in a network of computers connected to the internet. Some blockchain applications reside in a Cloud environment. Each computer node, a client, performs validation and relays the transaction to the next node. These computer nodes store a copy of the blockchain they validate.

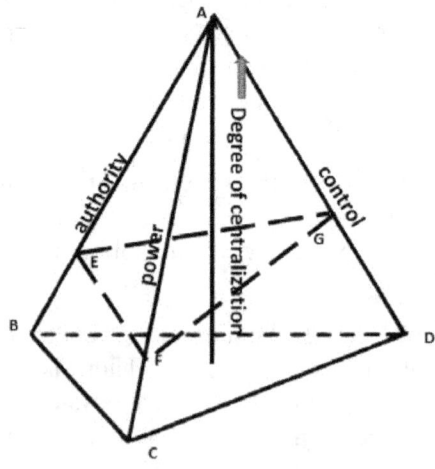

Figure 1-2 Degree of centralization

The chain can be as long as the history of transactions of a particular asset it represents. In the case of Bitcoin blockchain, it creates a block every 10 minutes and grows 6 blocks per hour, 144 blocks per day or 52,560 blocks per year. It contains all the transaction history of the Bitcoins. In addition, it exists in all the nodes in the Bitcoin blockchain network, it is impossible to alter the data and make them consistent throughout the chain during the ten minutes interval. The 10-minutes period is a result of the Proof of Work (PoW) algorithm, which takes miners 10 minutes to construct a block with hash lower than the target value.

Permissionless blockchain network is a subset of the internet. Every node is equal and joins the network voluntarily. This is a form of mass collaboration. The incentive for participating in the network as a miner is a chance of winning Bitcoins. In addition, the incentive for participating in the network as a full node is to have the most private and secure way to use Bitcoin.

Blockchain technology creates a channel for the peer-to-peer transaction and may change the way we do things at every level. The technology provides a distributed transaction ledger that connects each transaction to its previous transactions, protecting them all with encryption. The blockchain tracks and verifies digital assets so they cannot be hacked or copied without permission. It operates without the need for intermediaries. The high level of security is why blockchain is used for the cryptocurrency, but it can also be used for other assets, anything from votes in an election to stocks, tax payments, and property deeds.

What is revolutionary about the blockchain is that it decentralizes ownership and control of assets, and by doing this, it remodels fundamental structures in our society. It also replaces the traditional trust wholly or partially by cryptographic technology. By doing so, it removes one of the biggest barriers to transactions of the valuables. From this point of view, it is the most important invention since the paper money. It will stimulate the transactions or economic activities between unknown parties, which is previously impossible without the third party functioning as trust element. Potentially, the blockchain technology can unleash huge potentials of the untapped economic activities due to the lack of trust.

Our global economy is based on the power and trust we place — because we have no alternative — in intermediaries like banks, governments, utilities, and large internet companies like

What is blockchain?

Facebook and Google. They add no value to transactions beyond the trust, yet they wield tremendous power over us all, and make money from the transactions and slow them down in the process. If transactions and assets remain secure without these intermediaries, then we do not need them; we can retain our own property and stop paying them to insert themselves into processes.

Interestingly, blockchain will take power and wealth away from even relatively new corporate intermediaries like Uber and Facebook. If, for example with Uber, drivers and passengers can contract themselves securely without risking their identities or financial information, then the intermediary (Uber) is extraneous.

However, the Permissionless blockchain has its weakness. Because there is no central authority, any change requires majority consensus. Sometimes, the consensus fails because of different interests that different groups in the blockchain community represent. This is apparent in the recent disparity in fixing the Bitcoin scaling issue, that the miners are pitted against users. It ended up with a split of the Bitcoin blockchain.

There are many potential applications of the blockchain technology. Since it is a transaction-based system, the most obvious applications are in the finance. For example, international remittances, which have over US$400 billion transaction amount per year, are one of the primary targets of the Blockchain application. It not only cuts out the intermediary to handle the transaction, thus, lowering the cost, but is fast and secure.

Blockchain technology is secure because it is encrypted and its network does not have the vulnerability of the single system that computer hackers can exploit. Perceptibly, most of the service industries, such as banking, real estate agency, insurance, etc. will be transformed by the blockchain technology beyond recognition.

Banks will need to add value rather than simply guard money and move it around, because, with blockchain technology, buyers and sellers will be able to verify the presence of funds and exchange them on their own.

Utility companies will also change as individuals switch to highly efficient renewable sources and store extra power; they can then sell their reserve power locally, effectively creating multiple mini-grids — unless utility companies include consumers as partners.

Intellectual property is another area that blockchain can affect. Musicians can use blockchain to cut record labels, retaining their ownership rights while distributing their music and being paid directly by users. Similarly, journalists and writers can put publishers out of the loop, and sell their own work to readers directly. Ultimately, paying artists directly will better compensate the artists for their creative work because no middleman will take his cut. These applications are in the arena of the Permissioned blockchain, which is more centralized and owned by an entity or a company.

Even governments might eventually have a less involved role and more streamlined infrastructure with the help of blockchain. While societies would still probably opt for group decision-making and policymaking processes, with blockchain, the execution of programs and distribution of benefits could be managed efficiently without as much corruption.

When new technologies like self-driving cars come onto the scene, blockchain technology can help to ensure safety. The blockchain technology can also reduce the need for government regulations by making data and software available to all. It will even democratize access to space-based resources such as satellites.

Section 1.04   Bitcoin

The concept of blockchain evolves from cryptography. Cryptography is a very specialized discipline in computer science that uses either encryption technology to ensure the security of data stored or transferred. It is not widely used until last decade when the computing power and internet connectivity became faster and cheaply available.

In 2008, a researcher named Satoshi Nakamoto wrote a paper to propose the use of cryptography to create a digital

What is blockchain?

currency called Bitcoin.[10] In his original publication, he envisioned a purely peer-to-peer version of electronic cash that allows sending online payment directly from one party to another over the internet without going through a financial institution.

In building the Bitcoin platform, he embodied several elements including digital signature, network timestamp and hashing transactions into an ongoing chain of hashes. Such a chain forms a record that is unchangeable without redoing the Proof-of-Work. The basic concept of the blockchain technology was born. As long as a majority of owners in the nodes that do not collaborate to attack the network, the data are safe.

The network itself requires minimal structure. Network nodes broadcast messages on the best effort basis. They are free to leave and to rejoin the network at will. The functionality of the Bitcoin blockchain does not depend on a particular node. Such a flexible structure makes Bitcoin network easy to grow.

The blockchain encryption uses public and private "keys". The public key is users' address on the blockchain. It is a destination that people can send Bitcoins to across the network. The public and private key pair represent the ownership of the Bitcoins. The asset tied to the public key belongs to the owner of the public key. The private key is a password that gives its owner access to his digital assets.

Actually, the pair of public and private keys is one entity, because the private key can always generate the public key. Protecting the private key is essential to protect the digital assets associated with the private key. Therefore, safeguarding the private key is critical to the possession of the Bitcoins stored in the corresponding public key. Many forms of "wallet" can safeguard the private keys. Figure 1-3 shows Bitcoin public and private keys in printed form.[11]

Even without considering many possible applications of the blockchain technology, the technology itself is a core breakthrough in computer science; It is built upon more than 20 years of research

---

[10] "Bitcoin: A Peer to Peer Electronic Cash System": Satoshi Nakamoto white paper

[11] Generated from bitaddress.org

in cryptographic currency, and 40 years of research in cryptography, by thousands of researchers around the world. Finally, it took shape by Satoshi Nakamoto. He incorporated all the essential elements into one place: the Bitcoin software.

Bitcoin is a solution to a long-standing issue with digital cash: the double-spend problem. Digital file cannot have any value of an asset because anyone can copy or duplicate it with minimum effort. Satoshi Nakamoto solved the double spending problem in his original paper published in 2008.

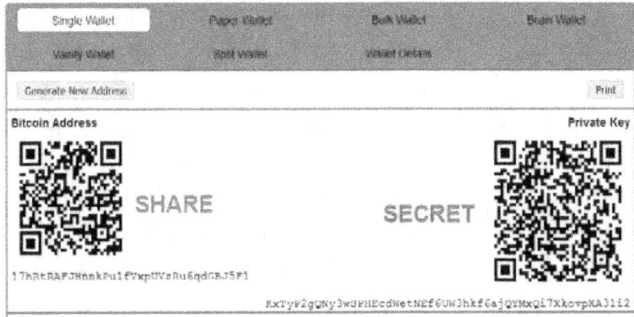

Figure 1-3 A printed Bitcoin Public Key and Private Key

He constructed the Bitcoin platform that hashes the transaction into an ongoing chain of hash-based Proof-of-Work, forming a record that cannot be changed without redoing the Proof-of-Work.

The ownership of Bitcoin is nothing more than the receipt of the transaction, which says you received a Bitcoin from the payer. The Bitcoin transaction is recorded in the blockchain and you can show your ownership by showing such a transaction by producing the Bitcoin address and password (private key). After you spend the Bitcoin in an address, your address will become invalid. You cannot use the same address again. This prevents the double-spending. The miners will check to see if the address is valid before accepting the transaction. If the address has been used in the previous transaction, it will be rejected. Bitcoin nodes doubly verify the validity of transaction before the block becomes official and attached to the blockchain. There is little chance that the Bitcoin in an address can be spent twice.

What is blockchain?

At the end of 2017, the price of Bitcoin reaches over US$18,000 at one time and the market cap of Bitcoin grew to well over US$180 billion as shown in Figure 1-4.[12] If we include the Bitcoin Cash, which is a fork of the Bitcoin, the combined market capitalization is even more. This is the proof that the concept of Bitcoin proposed by Satoshi Nakamoto is receiving enormous support.

Today, more and more businesses are accepting Bitcoin as payment. In Japan alone, over 260,000 stores accept Bitcoin as legal tender. In the US, big brand name companies are also accepting Bitcoins as payments, such as Whole Food, Expedia, Home Depot, CVS, Sears, Subway, Microsoft, among many others.

The main Bitcoin merchants processing solutions for vendors to accept Bitcoin are BitPay and Coinbase in the US and Coinify in Europe. Mobile payment functionality for quick point-of-sale Bitcoin purchases will further stimulate more widely adaption of the Bitcoins. There are already companies such as CoinBeyond developing mobile Bitcoin payment system.[13]

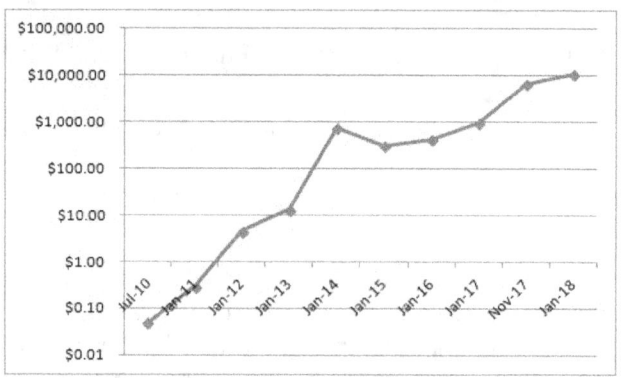

Figure 1-4 Year by year Bitcoin price

---

[12] https://bitinfocharts.com/

[13] "CoinBeyond Integrates with Bigcommerce to Enable 90,000 Merchants to Accept Bitcoins",CoinBeyond, https://www.coindesk.com/press-releases/coinbeyond-bigcommerce-merchants-accept-bitcoin/

Section 1.05 The proliferation of blockchain technology

By 2017, blockchain technology not only created many cryptocurrencies but also became the foundation of Fintech (the new financial technology). Hundreds of applications are flourishing. Many of them may make a huge impact in the near future.

With such explosive potential applications, blockchain technology enters the pivot year in 2017. The activity is not limited to the startups, a number of large companies, including financial, technological and logistic companies, are embracing the Blockchain technology.

Today, there are over thousand cryptocurrencies. Nevertheless, digital currency is not the only application of blockchain. Hundreds of transactional applications are in development using blockchain technology. Some people consider it the biggest invention since the invention of the internet itself.

The nascent blockchain industry is developing a new generation of transactional applications that establish trust, accountability, and transparency at their core while streamlining business processes and legal constraints. It is akin an operating system for marketplaces, data-sharing networks, micro-currencies, and decentralized digital communities. It has the potential to reduce the cost and complexity of the real world transactions.

For example, Linux Foundation hosts an Open Source, collaborative software development platform, known as Hyperledger, in conjunction with the leaders in finance, banking, internet of things, supply chains, manufacturing, and technology to develop the applications in their respective fields.[14] This approach ensures the transparency, longevity, interoperability, and support required to bring blockchain technologies to mainstream commercial adoption.

Likewise, R3, the largest blockchain consortium raised $107 million USD from global financial institutions like SBI

---

[14] https://www.hyperledger.org/community

What is blockchain?

Group, Bank of America Merrill Lynch, HSBC, Intel, ING, Banco Bradesco, Itaü Unibanco, Barclays, UBS, Wells Fargo and many others to develop commercial applications for the MDL technology.[15]

Once again, internet finds another value generation and revolutionary technological breakthrough since the invention of e-commerce (e.g. Amazon) and online advertisement through a search engine (Google). There is no reason to doubt that blockchain will not be able to give birth to giant companies like Amazon and Google. In fact, there is every reason to believe that blockchain technology will make an even bigger impact on the world than the search engine and e-commerce.

Not since the Web itself, a technology has promised broader and more fundamental revolution than blockchain.

Section 1.06   Initial Coin Offering (ICO)

Blockchain technology is creating a completely new financial market called Initial Coin Offering (ICO). ICO is a process that the cryptocurrency companies sell their newly launched crypto coins. ICO bypasses the regulated capital-raising process required by venture capitalists or banks. In an ICO campaign, the early backers of the project buy a percentage of the newly issued cryptocurrency. The proceeds are to fund the company. The ICO process is also known as crowdsales. Ethereum, a company that developed a blockchain protocol, which can transact smart contracts in a distributed network, is a great success story of Initial Coin Offering.

Much like the traditional angel funding, the startup company with a vision to create a business generates a project plan. The project plan includes a business plan with whitepaper. The plan describes the project mission and vision, detail plan, execution strategy, technological approach, target market, potential applications, goals, risks, and its differentiation from other similar projects, funding needed for execution, amount of new coins the founders keeping for themselves, and the ICO campaign period.

---

[15] https://www.r3.com/

The aim of the startup is to attract and convince enthusiasts and supporters to buy some of the distributed crypto coins. These coins are tokens. If the money raised does not meet the minimum required fund at the end of ICO campaign period, the ICO event fails. The startup returns the money to the subscribers of the ICO. Otherwise, the new coin is born and the money raised goes to fund the startup.

Early investors in the operation are usually motivated to buy the crypto coins in the hope that the plan becomes successful and the coins will increase value.

In the ICO model, the return on investment does not come from the shares of the company that they invest, but from the blockchain-based token. For example, an Ethereum token was worth of $1.00 on January 2016, it became $349.06 on June 10, 2017. It was a successful investment in a very short period.

The token represents an ownership of the blockchain network of the proposed project, not the company. This is a radically new way of investment. The ownership is the direct asset rather than the company holding the asset. Eventually, if the project becomes valuable, the token value will increase. Of course, if the project fails, the coin will become worthless.

Today, the blockchain technology is still in the incubation stage. Many projects may fail. When the real killer application appears, its growth will be explosive. In the process, it can create many millionaires. However, like any investment, investors need to do due diligence to understand the underlying technology to know whether it has any potential. Today, because the financial authorities do not yet regulate the ICO process, its risk is higher.

Many blockchain companies that have done ICOs have found major success, raising tens of millions of dollars at a massive valuation. In fact, the blockchain startup companies have raised more capital from ICOs than traditional early-stage and venture capital funding.

Instead of building a new blockchain from scratch, one can also build projects on the existing, open blockchain platforms. Over the past year, an increasing number of blockchain projects have utilized the Ethereum protocol to distribute unique tokens in a decentralized and transparent manner. The tokens released by independent Blockchain projects and companies are compatible with the Ethereum network and its native token.

Section 1.07   Blockchain platforms

Bitcoin blockchain is the first blockchain platform, but by no means the only one. In the last decade, many blockchain platforms have been developed. Each one has its own characteristics and uniqueness. They all target for certain applications. The technology has applications in sectors ranging from capital markets and trade finance, healthcare, and energy, to government taxation.

For instance, the technology could potentially be applied to capital markets to eliminate the need for reconciling various ledgers by providing a shared and synchronized blockchain among participants. We will discuss some of the most important ones.

For example, Chain Core is a blockchain platform for issuing and transferring financial assets on a Permissioned blockchain infrastructure. Corda is a distributed ledger platform with pluggable consensus. Credits is a development framework for building Permissioned distributed ledgers. Domus Tower Blockchain aims for regulated environments, benchmarked at handling over one million transactions per second. Ethereum is a decentralized platform that runs smart contracts on a custom built blockchain. HydraChain is an Ethereum extension for creating Permissioned Distributed Ledgers for private and consortium chains. Hyperledger Fabric supports the use of one or more networks, each managing different assets, agreements and transactions between different nodes. Hyperledger Iroha is a simple and modularized distributed ledger system with emphasis on mobile application development. Hyperledger Sawtooth Lake is a modular blockchain suite in which transaction business logic is decoupled from the consensus layer. The list goes on and on.

Moody published a report that identified over 120 on-going projects among the issuers that it rates. In the report, Moody evaluated the platforms in development by their efficiency in terms of speed, cost, security, and reliability, as well as the audit ability of processes. The longer-term viability of these platforms depends on whether their positives by leveraging blockchain technology will outweigh their negatives.

For the blockchain technology to make an impact on the existing establishment, there are many challenges. Among the challenges are the compatibility of blockchain technology with the existing systems, the need for the development of industry standards and questions around the regulatory treatment of the technology. The major consideration is that how the service can run smoothly and be transparent to the users when an existing system upgrades its service to a blockchain based system.

Many of the blockchain platforms in development are trying to address these challenges. Instead of seeing blockchain technology toppling the existing establishment, most likely it will greatly improve the efficiency and its cost.

Chapter 2   Bitcoin

Bitcoin is the first application of blockchain technology. We can also say that blockchain was invented to suit the Bitcoin application. Much of the other digital currencies and applications are variations of Bitcoin application. By understanding the inner works of Bitcoin blockchain, it is easier to understand other applications of the blockchain technology including non-cryptocurrency applications.

The objective of Bitcoin platform is to address the root problem of the fiat currency: all the trust that is required to make it work. Bitcoin platform replaces the trust with the use of cryptographic proof and decentralized networks.

Wei Dai at Microsoft published the earliest proposal for an "anonymous, distributed electronic cash system" in November 1998. Wei Dai called it the B-Money.[16] In his proposal, Wei Dai has first described the Proof-of-Work concept using Hashcash algorithm, although he did not use the term "Proof-of-Work". Wei Dai's proposal advanced three key concepts, which became the cornerstone of the blockchain technology:

- Transaction: Funds are exchanged through broadcasting and digital signature.
- Verification: The transaction is verified by authentication with cryptographic hashes, and updated to the collective ledger.
- Reward: Whoever does the work is rewarded for his effort.

---

[16] Wei Dai (1998) "B-Money", http://www.weidai.com/bmoney.txt

Bitcoin

Wei Dai did not put his idea into practice. His idea remained as an academic curiosity. Ten years later, Satoshi Nakamoto took the B-money concept further when he created Bitcoin platform. He first published the Bitcoin software at Bitcoin Core website in 2009 available to the public.[17] It is the original Bitcoin reference client or the Satoshi client.[18] The Bitcoin nodes use the Core software to create the Bitcoin network. Over the course of the following nine years, Bitcoin Core has served as the main branch of Bitcoin development.

After the disappearance of Satoshi Nakamoto in mid-2010, Bitcoin developers continue to support the Bitcoin Core website and make changes to the underlying Bitcoin protocol. Since 2011, it has released 44 versions of the Bitcoin software. To keep the promise of decentralization, Bitcoin Core does not take a strong position in the implementation of any modification to the Bitcoin software. It is up to the Bitcoin community to decide. Bitcoin Core also provides a wallet service for free download.[19]

The first block on the Bitcoin blockchain, known as block 0 or the Genesis Block, was mined on January 3rd, 2009, according to the timestamp in the block header, Figure 2-1 Bitcoin genesis block. Nakamoto made the Bitcoin software available online on January 8, 2009. The second user identified as cryptographer Hal Finney, a Caltech educated computer scientist, mined a few days later. Hal Finney himself has also advanced the Proof-of-Work idea in this publication in 2007.[20]

Because Bitcoin platform is distributed and decentralized, even though Bitcoin Core created the system, it does not own the system. As a reference client of Bitcoin, Bitcoin Core cannot enforce the changes to the Bitcoin software. Rather, the Bitcoin community has to reach a consensus on critical decisions. Such a decision by consensus creates difficulty to reach sometimes, because of the conflict of interest from different parties involved

---

[17] Satoshi Nakamoto, Bitcoin: A Peer-to-Peer Electronic Cash System. https://bitcoin.org/bitcoin.pdf
[18] https://bitcoin.org/en/bitcoin-core/
[19] https://bitcoin.org/en/download
[20] https://web.archive.org/web/20071222072154/http://rpow.net/

Bitcoin, Blockchain & Fintech

in the Bitcoin community. It is the inherent shortcoming of the leaderless state of Bitcoin platform.

Raw block data

The raw hex version of the Genesis block looks like:

```
00000000   01 00 00 00 00 00 00 00  00 00 00 00 00 00 00 00   ................
00000010   00 00 00 00 00 00 00 00  00 00 00 00 00 00 00 00   ................
00000020   00 00 00 00 3B A3 ED FD  7A 7B 12 B2 7A C7 2C 3E   ....;£íý z{.²zÇ,>
00000030   67 76 8F 61 7F C8 1B C3  88 8A 51 32 3A 9F B8 AA   gv.a.È.Ã^SQ2:Ÿ ª
00000040   4B 1E 5E 4A 29 AB 5F 49  FF FF 00 1D 1D AC 2B 7C   K.^J)«_Iÿÿ...¬+|
00000050   01 01 00 00 00 01 00 00  00 00 00 00 00 00 00 00   ................
00000060   00 00 00 00 00 00 00 00  00 00 00 00 00 00 00 00   ................
00000070   00 00 00 00 00 00 FF FF  FF FF 4D 04 FF FF 00 1D   ......ÿÿÿÿM.ÿÿ..
00000080   01 04 45 54 68 65 20 54  69 6D 65 73 20 30 33 2F   ..EThe Times 03/
00000090   4A 61 6E 2F 32 30 30 39  20 43 68 61 6E 63 65 6C   Jan/2009 Chancel
000000A0   6C 6F 72 20 6F 6E 20 62  72 69 6E 6B 20 6F 66 20   lor on brink of
000000B0   73 65 63 6F 6E 64 20 62  61 69 6C 6F 75 74 20 66   second bailout f
000000C0   6F 72 20 62 61 6E 6B 73  FF FF FF FF 01 00 F2 05   or banksÿÿÿÿ..ò.
000000D0   2A 01 00 00 00 43 41 04  67 8A FD B0 FE 55 48 27   *....CA.gŠý°þUH'
000000E0   19 67 F1 A6 71 30 B7 10  5C D6 A8 28 E0 39 09 A6   .gñ¦q0·.\Ö¨(à9.¦
000000F0   79 62 E0 EA 1F 61 DE B6  49 F6 BC 3F 4C EF 38 C4   ybàê.aÞ¶Iö¼?Lï8Ä
00000100   F3 55 04 E5 1E C1 12 DE  5C 38 4D F7 BA 0B 8D 57   óU.å.Á.Þ\8M÷º..W
00000110   8A 4C 70 2B 6B F1 1D 5F  AC 00 00 00 00           ŠLp+kñ._¬....
```

Figure 2-1 Bitcoin genesis block

Section 2.01 Bitcoin, a ledger

Bitcoin is a digital file that lists transactions like a ledger, Figure 2-2 A ledger. Bitcoin blockchain bears many similarities to the paper ledger. In a paper ledger, each page can record many transactions. Likewise, in a blockchain, each block can record many transactions. Each page in the ledger is bonded to the next page. The balance at the end of each page carries to the next page, thus forming a link between pages. In the same manner, each block of Bitcoin is bonded to the previous block to form a chain.

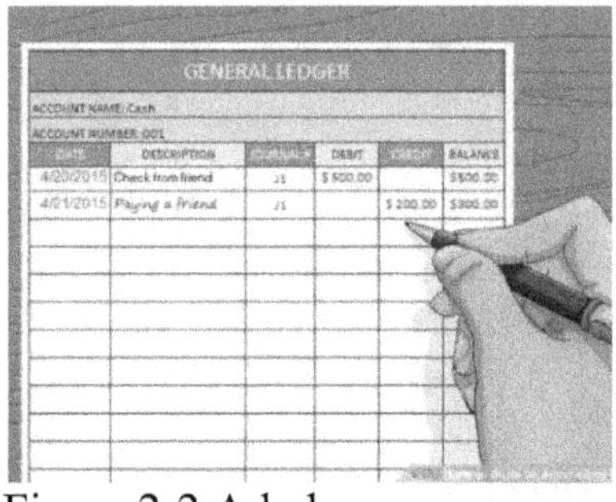

Figure 2-2 A ledger

Bitcoin

Every node in the Bitcoin network keeps an updated copy of the Bitcoin blockchain, which contains a complete transaction history of the Bitcoin. New transactions are broadcasted to the network. A special kind of nodes called the miners in the Bitcoin network collect the new transactions validate and record them in a new block. In the Bitcoin platform, the transaction history is transparent. The Bitcoin digital file contains a complete history of the Bitcoin transactions since its creation. Therefore, it makes the verification possible.

One of the differences between Bitcoin ledger and the traditional paper ledger is that all the nodes in the Bitcoin network maintain a copy of Bitcoin ledger, while the person that handles the transaction maintains the paper ledger.

Bitcoin platform must work without the need for trust because in a system with many anonymous users, trust does not exist. Bitcoin platform uses special mathematical functions to produce a secret code to replace the trust. These functions are to make the fraud impossible.

Bitcoin digital file, or Bitcoin blockchain, consists of blocks containing a sequence of digitally signed transactions that began with the Bitcoin's creation. Each Bitcoin in existence was created by a miner, who receives Bitcoins as a reward for doing the work of conducting the transactions per a set of rules of the Bitcoin platform. It is like that the accountant who maintains the paper ledger earns money by entering a transaction in the ledger page he has just completed to receive a payment for his service.

The payer of a Bitcoin transfers it by digitally signing it over to the payee using a Bitcoin transaction, much like endorsing a check.

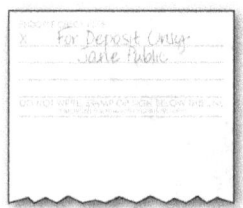

Figure 2-3 An endorsed check

Figure 2-3 shows a check endorsed and signed by the payer to the payee. In the Bitcoin platform, the process can continue to go on. That is the payee can, in turn, be a payer and endorse the check and pay to another person. A payee can examine each previous transaction to verify the chain of ownership. Bitcoin transactions are irreversible. Bitcoin is like money, which can change hands many more times than a check.

Section 2.02   How does Bitcoin work?

Bitcoin network is a peer-to-peer network; there is no central authority in charge of creating currency or verifying transactions. All the work is performed in the Bitcoin network, which has distributed computing power. Anyone can be part of the Bitcoin network by downloading and executing Bitcoin software.

There are three types of nodes in the Bitcoin network: miners, complete nodes, and partial nodes. Miners perform the work of transaction verification, encryption, and building the blocks. Other nodes have the responsibility of validating the blocks built by the miners. Complete nodes carry a complete copy of the blockchain, while partial nodes carry only part of the blockchain. Miners are rewarded for their work of building blocks with newly created Bitcoins, while other nodes are strictly volunteers. Currently, there are ~10,000 complete nodes in the Bitcoin network.

The Bitcoin software encrypts each transaction—the sender and the receiver are identified only by a string of numbers, their public keys, which are public information. Buyers and sellers remain anonymous, but everyone can see that a coin has moved from address A to B, and the code prevents A address from spending the coin again.

The central idea of Bitcoin is that it can be used as a currency. We can best understand the concept of Bitcoin transaction by studying the representation of transaction scheme in Satoshi Nakamoto's original paper.[21] (Figure 2-4)

---

[21] "Bitcoin: A Peer-to-Peer Electronic Cash System", Satoshi Nakamoto, https://bitcoin.org/bitcoin.pdf

Bitcoin

A transaction involves three elements: a record of the previous transaction, payer's private key and payee's public key. The private key is payer's evidence of ownership of Bitcoin. Since Bitcoin transaction record does not exist outside of the Bitcoin blockchain, the payee must have a new address or public key to receive the Bitcoin. The public key can be created by an online wallet. Once these three elements are in place, the transaction is broadcasted to the Bitcoin network.

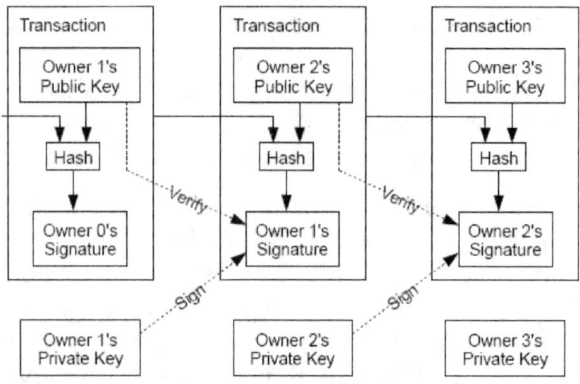

Figure 2-4 Satoshi's illustration of Bitcoin transaction in this original publication[2]

Miners collect these transactions and validate the data in the transaction. One of the most important validations that miners check is the validity of the payer's public key. This is because if the public key has been used in a previous transaction, it is invalid. An invalid public key make an invalid transaction. Miner must reject such a transaction. This solves the so-called double spending issue. It guarantees that no Bitcoin can be used in two different transactions. After validation, the miner then produces payer's digital signature. We will discuss in the next section how the digital signature is done.

All the hashes are done by an algorithm called SHA-256. SHA256 is a version of the Secure Hash Algorithm. It generates 32-byte or 256-bit hash code. The possible combinations of such hash code are in the order of 1×10^{77}, roughly the order of number of all the atoms in the universe, which is infinite in practical purpose. Therefore, the hash can be considered as collision-free. It is nearly impossible to find two input strings with the same hash. As such, the hash of a data string is "unique". More discussion on hash in Section 2.04.

Section 2.03   Digital signature

Digital signing is a process that fuses the message (transaction content) with the identity of the owner (private key) mathematically. It produces an output that contains both the message and the identity of the owner, and others can verify the signature but not alter it. The process is much like signing a document that your signature, representing you, fuses into the document. Your signature in the document can be verified but not altered.

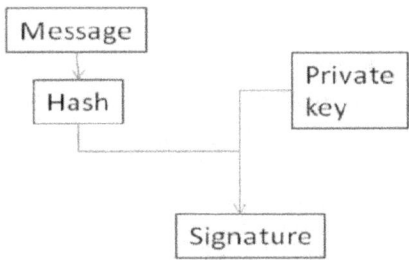

Figure 2-5 Digital signature flow

A digital signature executes three functions:

- Authentication, that the message was created by a known sender,
- Non-repudiation, that the sender cannot deny having sent the message,
- Ensuring integrity, that the message was not altered in transit.

To sign digitally, you use a private key that is unknown to anybody else. However, your digital signature can be verified by using your public key. A signature is generated from a hash of the message (e.g. the transaction data) to be signed, plus a private key (Figure 2-5).

Bitcoin

To verify the signature, the Bitcoin software uses an algorithm and the signer's public key as input to determine if it was originally produced from the hash and the private key. It does not need the private key of the signer. (Figure 2-6)

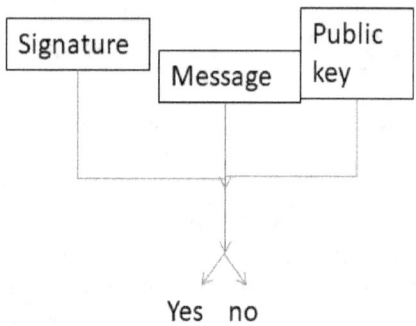

Figure 2-6 Digital signature verification

Bitcoin software ensures that all nodes in the Bitcoin network have the record of all transactions in the Bitcoin network constantly updated and verified. This prevents double spending and fraud.

There are many digital signature algorithms. The most popular ones are RSA (Rivest, Shamir, Adleman), DSS (Digital Signature Standard). Bitcoin platform uses the algorithm called ECDSA or Elliptic Curve Digital Signature Algorithm. Bitcoin platform chose ECDSA because it produces signatures and public keys much smaller than RSA or DSS at similar security levels. The RSA signature and public key are 128 bytes each as compared to 48 bytes-long signature and 25 bytes-long public key produced by ECDSA. Even so, the signature is a large portion of the transaction data size.

In addition, the generation of signature depends on the software that combines the private key with other data. Software library such as "OpenSSL" is for such a purpose. In order to decrease the validation time, Bitcoin Core developers developed a new and more efficient library called "libsecp256k1" to replace the "OpenSSL" originally used for validation. The signature validation using "libsecp256k1" is 2 to 5 times faster.

The large data size of the signatures also consumes precious block size. Fewer transactions can fit into a block when the signature data size is larger. Since it takes 10 minutes to produce a block, and currently, each block can contain no more than 3,000 transactions, this limits the transaction rate of Bitcoin platform by 3,000 transactions per 10 minutes, which is far from sufficient when Bitcoin becomes popular worldwide.

One of the proposals to reduce the transaction data size and therefore increase the transaction rate of the Bitcoin platform is to separate the signature from the transaction. This proposal is called Segregated Witness or SegWit in short.

There is also a different kind of signature scheme called multisig, or multisignature. In a multisig case, more than one key or signature is required to authorize a transaction. In the real world, there is the need that two or more persons must agree in order to spend the fund. You can do this by creating a multisig wallet. The transaction is sent to a multisig address.

The multisig scheme is an extension of the P2SH (Pay-to-script-hash). The script defines who needs to sign to spend the Bitcoin at this address. In other words, the fund sent to the P2SH address with script specifying the private keys to unlock the fund will require these private keys to sign in order to spend it.

A multisig scheme can be used for a single signer to enhance the security. For example, the Bitcoin can be deposited in an address which requires two keys, one kept by the user and one kept by the exchange. (Figure 2-7) Both keys need to be present to spend the fund.

The multisig scheme is getting more popular. The number of multisig address has reached 1 multisig output per block recently from almost nothing a few years ago. Granted, it is still a very small number, but it is increasing exponentially, thanks to the wider adaption of the multisig wallet.[22]

The efficiency of generating signature has an impact on the Bitcoin mining. Signature is a unique string of numbers generated

---

[22] "The year of multisig: How is it doing so far?" by Thomas Kerin, Coindesk.

Bitcoin

by the private key with other input data using cryptographic software. The corresponding public key can verify that the signature was created using the private key. Since the private key is embedded in the signature, it effectively proves the ownership of Bitcoin and validates the transaction.

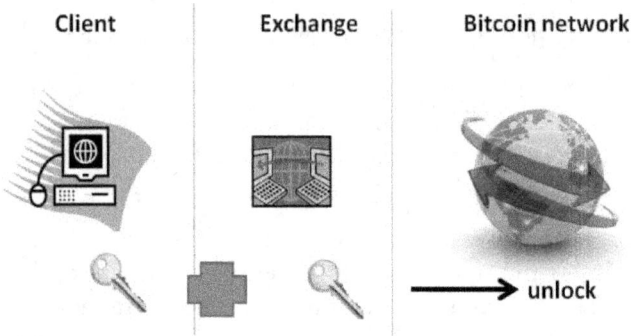

Figure 2-7 Multisig for enhanced security

Section 2.04   Hash

Hash is a cryptographic function. Once the miner validates the transaction, it time-stamps and produce a hash of the transaction using SHA-256 algorithm.

The idea of hash was first proposed by C. Dwork and M. Naor in 1992. It was Adam Back who invented the Hashcash algorithm in 1997. Hashcash is a cryptographic hash-based algorithm that requires a selectable amount of work to compute. It was originally developed for screening the spam email. In the email system, a textual encoding of a hashcash stamp, characterized by the amount of time that sender spends to prepare the email, is added to the header of an email. A spam email is likely to be prepared by an algorithm, therefore, has a very different stamp. By using Hashcash, it is easy to tell email composed by human or by spam algorithm. Microsoft used the Hashcash algorithm for email spam prevention in Hotmail, Exchange, and Outlook.

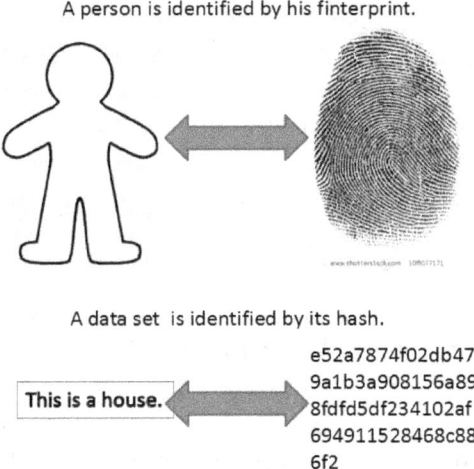

Figure 2-8 Hash and fingerprint

The purpose of hashing is to produce a unique, reproducible and yet simple data string which represents a record or dataset. A hash can identify the integrity of a dataset, but it cannot reconstruct the original dataset. Hash to data is like fingerprint to a person (Figure 2-8). Any change in the input data, even slightly, will generate a different hash. Since hash is a simple data string, it can be verified easily. The data is thus temper-proof. This is essential to many blockchain applications.

Let us use a very simple hash function to illustrate the concept of hashing. For example, we want to produce a hash 30 bytes long of a document (e.g. contract) which contains 15,630 characters. We divide 15,630 by 30, and get 521. We then select every 521st letter of this document and form a string of 30 bytes long. This hash uniquely represents the original document since any alteration in the document will produce a different hash. Of course, this hash function is too simple. If we change a letter not included in the hash, it will produce the same hash and yet the dataset is different. This is called collision.

Let us look at another hashing example: To take the square root of a number, say, 3.

$$\sqrt{3} = 1.73205080756888$$

Bitcoin

We then take the string of numbers from the fifth to ninth digits after decimal point: 50807. We call 50807 as the hash of 3. It is a unique representation of 3. We thus have a simple hash function which consists of two steps: 1) the square root, 2) the string number from the fifth to ninth digits after decimal point of the square root number.

If we change 3 to 3.0001, the hash changes to 79674.

$$\sqrt{3.0001} = 1.73207967484178$$

In this simple hash example, we can produce a hash of any positive number, but not negative number or text.

The hash algorithm used in the Bitcoin platform is the SHA-256 (Secured Hash Algorithm), which includes checksum, randomization, error-correction, ciphers, and cryptographic functions.

Historically, the first SHA, SHA-1, was developed in 1993. It produces 20-byte hash value. Over time, improved version of SHA, such as SHA-2, and SHA-3 were developed. SHA-256 used in the Bitcoin blockchain is much more secure and sophisticated.

You can experiment the SHA-256 hashing function by using a demo program at the Blockchain Demo site. [23]

We used the demo SHA-256 hash program on-line to generate hashes from three data as listed in Figure 2-9 Examples of hashing using the demo program. We produced three hashes using short data, medium length data, and long data. The hash strings are generated in the box below the data. Independent of the data size, all hash strings are of the same size. You can hash a data set many times. As long as the data are the same, the hash string is always the same. In the bottom box, we can clearly see that the hash is much shorter than the input data. This illustrates the point that hash does not contain enough information that can restore the original input data.

---

[23] https://anders.com/blockchain/hash.html.

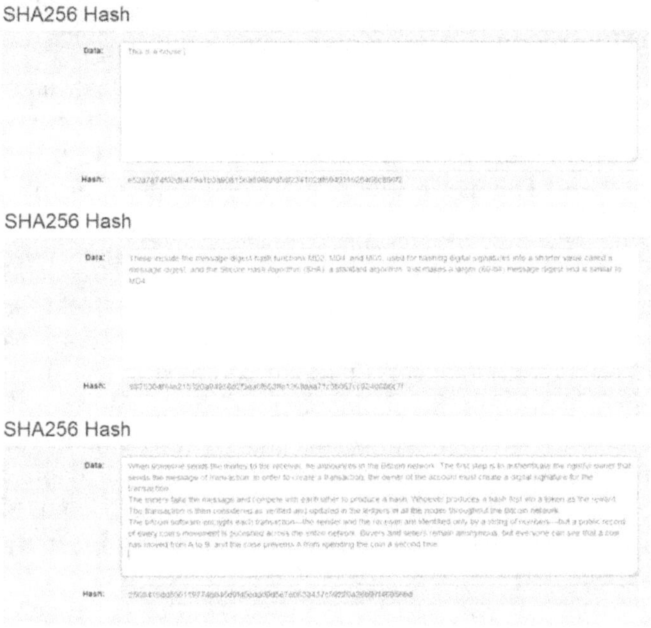

Bitcoin, Blockchain & Fintech

Figure 2-9 Examples of hashing using the demo program

In the demo program, also try to add a space anywhere in the paragraph or even at the end of a paragraph that you do not see, you will get a completely different hash result, but the hash is of the same length. This is the beauty of hashing. You can hash one paragraph, one chapter or one book, the size of the hash is always the same. If you use the same algorithm and input the same data, you will get an identical hash string. You can hash text as well as a picture or any digital data.

More discussion of SHA-256 algorithm can be found at the reference.[24] Satoshi Nakamoto referred to both Wei Dai and Adam Back in his white paper.

Hash is like a DNA of a Bitcoin. Transaction 1's hash is included in the data that produce Transaction 2's hash. In such a way, the two transactions are intimately linked. Altering Transaction 1 will inevitably alter Transaction 2 as well so that the

---

[24] "SHA-256 Cryptographic Hash Algorithm", Chris Veness, https://www.movable-type.co.uk/scripts/sha256.html

Bitcoin

Proof of Work can verify the history of the Bitcoin involved in the transaction.

A hash, generated using SHA-256 on the header of the block, represents a block. This hash embeds itself inside of the header of next block in the blockchain. In this way, the blocks form a chain. This is as if a child's DNA contains his/her parents DNA.

SHA-256 is also used for other crypto coins: [25], [26] NameCoin (NMC), Devcoin (DVC), IxCoin (IXC). So far, SHA-256 is the most widely used hash algorithm for the cryptocurrencies. However, there are cryptocurrencies using other hash algorithms. For example, "Scrypt" hash algorithm[27] is used in forLitecoin (LTC), DogeCoin (DOGE), FeatherCoin (FTC), WorldCoin (WDC), Reddcoin (RDD). "Scrypt Adaptive-N" is used for Vertcoin (VTC), ExeCoin (EXE), GPUcoin (GPUC), ParallaxCoin (PLX), SiliconValleyCoin (XSV). "Scrypt-Jane (Scrypt-Chacha)" is used for YaCoin (YAC), Ultracoin (UTC), Velocitycoin (VEL). SHA-3 (Keccak) is used for MaxCoin (MAX), Slothcoin (SLOTH), Cryptometh (METH), etc.

Section 2.05   Merkle tree and block header

The miner takes hashes of two transactions, say hash(tx1) and hash(tx2), and hash them again into a new hash called hash[hash(tx1),hash(tx2)], and repeat the process for more transactions. The result is a tree of hashes called the Merkle tree. (Figure 2-10) [28] In such a process, miner builds pyramid-like hash structure for all the transactions that can be fit into a block, roughly 3,000 transactions. Eventually, all the hashes merge into

---

[25] "List of SHA-256 coins",
https://bitcointalk.org/index.php?topic=482002.0
[26] "SHA-256 coins", Bitcoinnational, Steemit.,
https://steemit.com/sha256/@bitcoinnational/sha-256-coins
[27] "Cryptocurrencies mining on Scrypt algorithm",
https://bitmakler.net/scrypt___mining
[28] "Merkle tree Bitcoin", Crudden Research Group,
http://simlamagto.cakedecoratingpro.info/jixuj/merkle-tree-bitcoin-3153.php

one hash, called the root hash, which is referenced into the block's header. This is all done within ten minutes interval.

The overflow transactions will form a queue and stay in Miner's memory pool. Miners prefer to pick up the transactions that pay higher fees. If there are too many overflow transactions in the pool, those transactions paying lower fees could stay in the pool for a long time without being incorporated into blockchain.

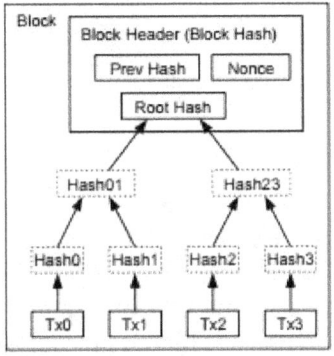 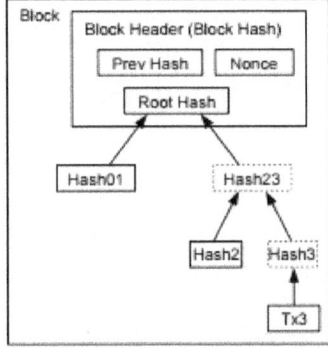

Transactions Hashed in a Merkle Tree   After Pruning Tx0-2 from the Block

Figure 2-10 Merkle tree

The root hash contains the hashes of all the transactions. Because any change in one of these transactions built into the Merkle tree will alter the root. The intermediate hashes do not need to be stored. This greatly saves the disk space in miner's nodes.

After having a valid Merkle root, the miners build the block's header. The header includes a hash of the previous block, thus making a chain of blocks, Merkle root, Timestamp, a representation of the networks current difficulty, and finally, the Nonce, a number incremented when mining.

Block header contains all the essential information about the block. The structure of header is shown in Figure 2-11.

Bitcoin

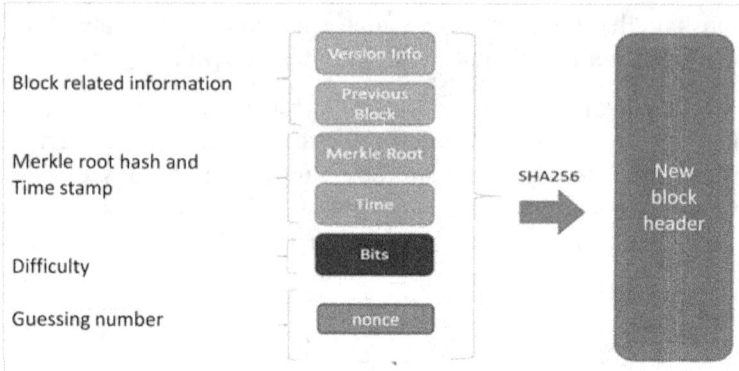

Figure 2-11 Block header

Even though SHA-256 is complex to generate, alias, the modern computers are too powerful to produce an SHA-256 hash in a fraction of second. If hashing is done by a simple SHA-256 calculation, a block can be constructed in a fraction of second. It would make the most powerful computer in the Bitcoin network to dominate the building of blockchain completely. Such a system would not be decentralized. To address this issue, Satoshi introduced another concept: "difficulty and nonce".

Section 2.06   Difficulty and nonce

The purpose of introducing difficulty is to increase the difficulty for producing the required hash. It is doing so by demanding that the hash produced must meet certain criterion. This requires multiple iterations to produce hash, therefore, increasing the difficulty of obtaining a valid hash.

Not only it makes hash more difficult to produce, but it also acts as an equalizer that the miner with the highest hashing power will not always be the winner each time. This is because the nonce is random and the transactions picked up by the miners are also random. A miner having more hashing power may randomly encounter the situation that he will have to do more work. In the end, the miner will build more blocks in proportion to its hashing power. The miner with the highest hashing power will not always be the winner.

Nakamoto proposed to do this by:

1) Requiring the valid hash begin with certain number of zero's.
2) A nonce is incorporated into the message. If the hash does not meet the number of zero's requirement, continue the hashing process by changing nonce number.

The following example illustrates the procedure. In this case, we require that the hash must start with a zero. If we produce the SHA-256 hash of "My name is John. 00000001", it is 841286639d940470cec3951322fd3502bf1a21488f28327bb5d15f044b5df8f7, which does not begin with a zero. We continue to hash by changing the nonce.

Table 2-1 Hashing with nonce

MESSAGE	NONCE	HASH
My name is John.	00000001	841286639d940470cec3951322fd3502bf1a21488f28327bb5d15f044b5df8f7
My name is John.	00000002	ccd546f9d30552cbe0cb2ba17f74faf99eda1f94e339b7d68b1c55622ce64f27
My name is John.	00000003	0abde8823f988ccdccf7db44d3f75466158b85143f2b6aa217b6f9801bf03d16

In the third trial, we obtain a hash beginning with a zero. If we require that a valid hash must begin with, say 3 zero's, it will take more trials. Therefore, the difficulty increases. When the difficulty increases, the demand of hashing power, or the computing power, increases proportionally. Since the transactions are randomly picked up by the miners, there is no telling what combinations of transactions will require less hash iterations. Thus, in average, the chance of producing a block will be proportional to the hashing power of the miner. The miner with most powerful hashing power will not always be the first to produce a block.

The difficulty is calculated from the formula:

$$Difficulty = \frac{genesis\ block\ hash\ value}{current\ target\ hash\ value}$$

In the example above, the difficulty is

$$Difficulty = \frac{841286639d940470cec3951322fd3502bf1a21488f28327bb5d15f044b5df8f7}{0abde8823f988ccdccf7db44d3f75466158b85143f2b6aa217b6f9801bf03d16}$$

Bitcoin

Apparently, the smaller the target value, the larger is the difficulty. An acceptable hash must be smaller than the target. It is a number beginning with a certain number of zeros. All Bitcoin clients share the same target value. The difficulty adjusts itself every 2,016 blocks, based on the network's recent performance, with the aim of keeping the average time between new blocks at ten minutes. When the hashrate of the Bitcoin network increases, the difficulty also increases. As of August 2019, the target hash value begins with 19 zeros.

The level of difficulty is ever increasing since the Bitcoin mining started. (Figure 2-12)[29]. In 2009, the difficulty was near 1. We also noted that the price of Bitcoin is in line with the difficulty. This is expected because if the difficulty increases much more than the price, miners will lose money by paying energy consumption more than the reward of Bitcoins they receive. This would discourage mining and deprive the construction of new blocks.

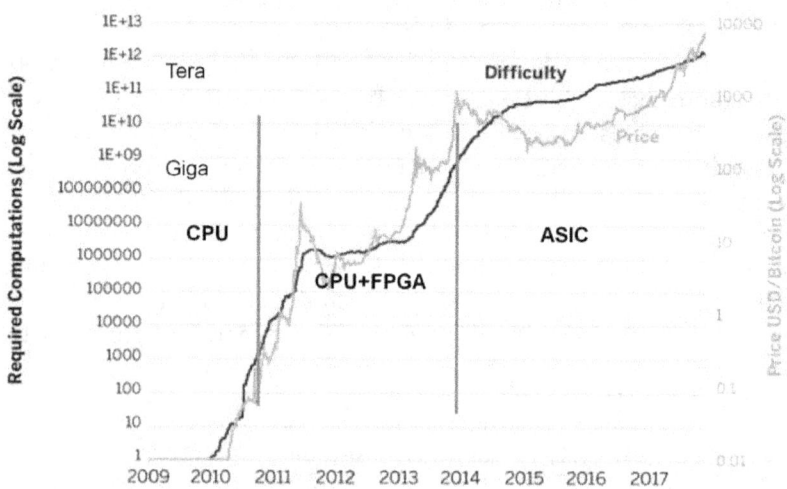

Figure 2-12 Difficulty of Bitcoin network

The work performed by miner is called Proof of Work. Proof of Work is a try and error process. When doing the Proof of Work, the miner receives a random nonce number between 0 and

---

[29] "A guide to Bitcoin (part 1): a look under the hood", Alex Barrera, http://tech.eu/features/808/bitcoin-part-one/

the maximum value of a 256-bit number. If a hash produced is below the target, the hash is complete. If not, the process repeats with the nonce increased by one, until the hash below the target is obtained. For every 2016 blocks generated, Bitcoin clients recalibrate the difficulty.

Readers can go to the Blockchain demo website[30] to mine a hash for any data to understand the concept of the nonce. As the difficulty increases, it requires more computer computing power to mine the Bitcoins or to do the Proof of Work. Running more powerful computers and longer time will consume more electricity, in addition to the depreciation of more expensive equipment, therefore, the cost of mining Bitcoins increases over time. Unless the value of Bitcoins can catch up with the cost of resources, it does not pay to mine Bitcoins anymore. Miners have the discretion of choosing transactions randomly from a pool of unprocessed transactions that were broadcasted to the Bitcoin network. To make up the higher cost of mining, miners seek to maximize their profits by processing transactions which pay higher fees.

Each block contains the hash of the preceding block, thus each block has a chain of blocks that together contain a large amount of work. Changing a block requires regenerating all successors and redoing the work they contain. This protects the blockchain from tampering.

SHA-256 is also used for other crypto coins: [31, 32] NameCoin (NMC), Devcoin (DVC), IxCoin (IXC). So far, SHA-256 is the most widely used hash algorithm for the cryptocurrencies. However, there are cryptocurrencies using other hash algorithms. For example, "Scrypt" hash algorithm[33] is used in forLitecoin (LTC), DogeCoin (DOGE), FeatherCoin (FTC),

---

[30] https://anders.com/blockchain/block.html and
http://bitcoin.sipa.be/
[31] "List of SHA-256 coins",
https://bitcointalk.org/index.php?topic=482002.0
[32] "SHA-256 coins", Bitcoinnational, Steemit.,
https://steemit.com/sha256/@bitcoinnational/sha-256-coins
[33] "Cryptocurrencies mining on Scrypt algorithm",
https://bitmakler.net/scrypt___mining

Bitcoin

WorldCoin (WDC), Reddcoin (RDD). "Scrypt Adaptive-N" is used for Vertcoin (VTC), ExeCoin (EXE), GPUcoin (GPUC), ParallaxCoin (PLX), SiliconValleyCoin (XSV). "Scrypt-Jane (Scrypt-Chacha)" is used for YaCoin (YAC), Ultracoin (UTC), Velocitycoin (VEL). SHA-3 (Keccak) is used for MaxCoin (MAX), Slothcoin (SLOTH), Cryptometh (METH), etc.

One can imagine that before a block is formed and added to the blockchain, there are many versions of the block, because the competing miners pick up random transactions from the pool of unverified transactions. This is called forking. Only the first block accepted by the consensus of nodes becomes the official block to be added to the blockchain. Forking disappears when a block is added to the blockchain.

Section 2.07   Bitcoin supply

One interesting aspect of the Bitcoin is that its supply is limited. The creation of new Bitcoin slows down approximately every four years. Eventually, the total Bitcoin supply will be limited to around 21 million Bitcoins. The limited supply of Bitcoin is built into its software by design.

Bitcoins are created every time when a new block is mined. The network allows a new block created every 10 minutes. So, every hour, 6 blocks are mined. During the first 210,000 blocks, miners were rewarded with 50 Bitcoins for each block mined, and the reward decreases by 50% for every 210,000 blocks mined thereafter, or approximately four years at the rate of 10 minutes per block. So, in the first 4-year cycle (from 2009 to 2012), 210,000 (blocks) times 50 Bitcoins are mined, or 10.5 million Bitcoins. For the next 210,000 blocks mined (2013 to 2016), 210,000 times 25 Bitcoins are mined, or 5.25 million Bitcoins. Every 210,000 blocks later (or roughly 4 years), the Bitcoins mined will be halved. Currently, in 2017, the reward for a block is only 12.5 Bitcoins. The result is that the total number of Bitcoins will not exceed 21 million. Figure 2-13 shows the Bitcoin supply vs. the total blocks mines, the time required, reward per block mined, etc.

The actual Bitcoins available are less because some Bitcoins are lost and destroyed. Bitcoins can be permanently lost by losing the private keys.[34] The private keys in a hard drive will be gone together with their Bitcoins if the hard drive crashes. This actually happened.[35] Therefore, it is advisable to create duplicated copies of the private keys, or even printed out on paper. Bitcoins can also be lost by sending to a bogus address.

total blocks mined	time required in years	reward per block	Bitcoins generated per period	total Bitcoins generated
210,000	4.0	50	10,500,000	10,500,000
420,000	8.0	25	5,250,000	15,750,000
630,000	12.0	12.5	2,625,000	18,375,000
840,000	16.0	6.25	1,312,500	19,687,500
1,050,000	20.0	3.125	656,250	20,343,750
1,260,000	24.0	1.5625	328,125	20,671,875
1,470,000	28.0	0.78125	164,063	20,835,938
1,680,000	32.0	0.390625	82,031	20,917,969
1,890,000	36.0	0.195313	41,016	20,958,984

Figure 2-13 Bitcoin supply

Sometimes, this is done intentionally, as in the case of Proof of Burn to create a counterparty currency. One may think that 21 million Bitcoins may not be enough to support the Bitcoin transactions in the future. Fortunately, Bitcoin is divisible. 1/1000th of a Bitcoin is 1 MBTC. Today, the smallest Bitcoin unit is Satoshi. A Satoshi is a one hundred millionth. The unit has been named in homage to the original creator of Bitcoin, Satoshi Nakamoto.

Section 2.08 P2PKH and P2SH Bitcoin addresses

---

[34] http://www.telegraph.co.uk/technology/news/11362827/The-625m-lost-forever-the-phenomenon-of-disappearing-Bitcoins.html
[35] https://www.theguardian.com/technology/2013/nov/27/hard-drive-bitcoin-landfill-site

Bitcoin

A Bitcoin address is a secure identifier to accept Bitcoin payments, just like an email address to receive emails. Since the address must be publicly known, for anyone who wants to send the funds to, it is also known as the Public Key. The owner of the address is the only one that can unlock the address to exclusively access the funds by using the private key associated with the address (the public key). This is akin to the email system that the owner must use a password to log into the account to see his emails.

Pay-To-Public-Key-Hash (P2PKH in short) is the standard format for Bitcoin address. A standard P2PKH address has 34 letters/numbers combination. You can create P2PKH address by using any wallet app. For example, in the Armory wallet, you get an address to receive Bitcoins as soon as you click the "receive Bitcoins" icon on the page. Only the person holding the private key can sign a transaction with cryptocurrency token assigned to this address – while everybody who knows your address can verify the validity of your signature. Figure 2-14 shows how to create a Bitcoin address in the Armory wallet.[36]

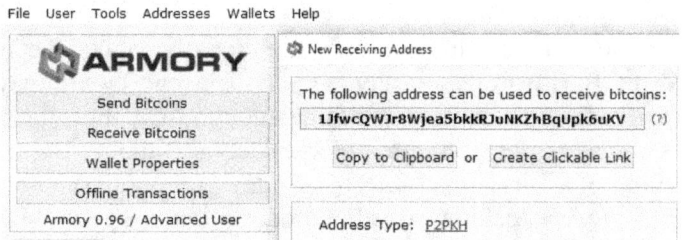

Figure 2-14 Armory wallet

A typical P2PKH address looks like 15Cytz9sHqeqtKCw2vnpEyNQ8teKtrTPjp. The hash starts with "1". To spend Bitcoins in a P2PKH address, the payer must use the private key associated with the public key hash to create a digital signature. The flow chart of the procedure is shown in Figure 2-15.

---

[36] https://www.bitcoinarmory.com/

Bitcoin, Blockchain & Fintech

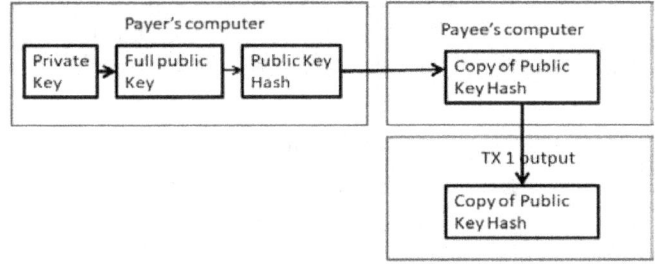

Figure 2-15 P2PHK payment flow

There is another type of Bitcoin address: P2SH or Pay-To-Script-Hash. A transaction using P2SH will send payment to a scriptPubKey instead of the public key address. A P2SH address

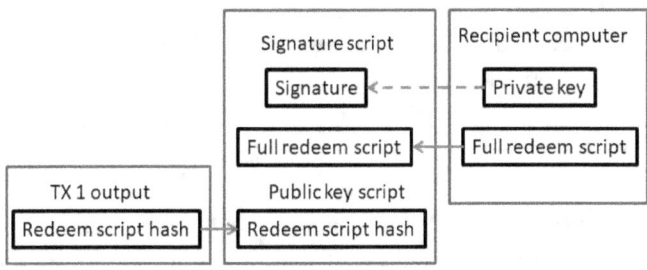

Figure 2-16 P2SH payment flow

always begins with a '3', instead of a '1' as in P2PKH addresses. The purpose of pay-to-script-hash is to move the responsibility for supplying the conditions to redeem a transaction from the sender of the funds to the redeemer.

The benefit is allowing a sender to fund any arbitrary transaction, no matter how complicated, using a shorter (20-byte) hash, which is more efficient in the data usage and results in lower transaction fees.

If you want to use P2SH address instead of the P2PKH address, you can do so by selecting the address type in the wallet. The scriptPubKey contains a script that is a set of instructions defined by the receiver describing how the receiver wants to gain access to the Bitcoins sent. For example, the instruction may ask the receiver to provide a public key that yields destination address

Bitcoin

embedded in the script, and a signature to show evidence of the private key. Or, the instruction may dictate who need to sign to spend this Bitcoin. The signature party can be one or more than one. We will discuss this in more detail in the next section.

You can imagine that payment by P2PKH is sent directly to the payee's public address. While in the payment of a P2SH scheme, payer sends the fund to a payee designated location, and payee retrieves the fund from such a location by using scriptPubKey.

To spend Bitcoins sent via P2PKH, the recipient must provide a Private Key to match the Public Key. On the other hand, to spend Bitcoins sent via P2SH, the recipient must provide a scriptPubKey (Figure 2-16)[37] instead of the public key. The scriptPubKey then reproduce a redeem script hash and checked against the redeem script hash in the signature. If they match, the transaction is verified. Pay-to-script-hash (P2SH) transactions were standardized on BIP 16.

Section 2.09   Zero Knowledge Proof (ZKP)

Zero Knowledge Proof means that one can prove something is true to others without giving away any information. An excellent example is Sudoku. You can prove to your friend that you successfully completed a Sudoku puzzle without revealing how you did it. Your friend will not learn from you how you did it. But he can verify that you did it correctly.

Another example, a magician shows you a trick by putting a rabbit in the box and proves to you that the rabbit is not in the box. You can say that magician's job is to perform ZKP.

In our daily life, we use ZKP methods all the time. For example, when you pay by credit card, the cashier asks you to input PIN to prove that you are the card owner. The cashier does not know your PIN because you do not reveal it to him/ her. But by doing so, you prove your identity.

---

[37]"Bitcoin-multisig-the-hard-way-understanding-raw-multisignature-bitcoin-transactions", http://www.soroushjp.com/2014/12/20/

In the ZKP protocol, the person who verifies challenges the person who proves. However, the challenge is in such a way that the person who verifies does not learn the knowledge required to prove it. Otherwise, it cannot be called a ZKP. Using ZKP, a person can verify the end result, even though he does not how to do it.

A ZKP program is a cryptographic system which lets a person run an arbitrary program with a mixture of public and secret inputs and prove to others that this specific program accepted the inputs, without revealing anything more about its operation or the secret inputs.

In the cryptocurrency and blockchain world, ZKP is very useful. It can be used in an authentication system where one party wants to prove its identity to a second party via some secret information but doesn't want the second party to learn anything about this secret. A ZKP can be as simple as a PIN, but it can also be very sophisticated to provide the extremely secure environment. [38]

In February 2016, the Bitcoin Core has integrated Zero-Knowledge Contingent Payment (ZKCP) on the Bitcoin network. This makes the Bitcoin transaction even more secure. ZKCP is a protocol that allows transactions be private, scalable and secure. The parties to the transaction do not need to trust each other or depend on arbitration by a third party.

Using a ZKCP avoids the significant transactional costs involved in a sale which can otherwise easily go wrong. Other cryptocurrencies are also either having ZKP authentication implemented or on their roadmap.

Section 2.10   Divisible Bitcoins

---

[38] "Efficient Zero-Knowledge Contingent Payments in Cryptocurrencies Without Scripts": Waclaw Banasik et. al. https://pdfs.semanticscholar.org/5cbf/e46f4b026f8dee4afb1e788236b3fdf08b81.pdf

Bitcoin

Each transaction has a different amount of Bitcoins. How do you divide the Bitcoins, if you have received 5 Bitcoins as one transaction in the past, but you want to spend 2 Bitcoins? Bitcoin resolves this issue by using multiple inputs and outputs, which allow Bitcoins to be split and combined.

Transactions can have multiple inputs combining their amounts, and multiple outputs: one for the payment and one return the change to the sender if any. The difference between the total input and output amounts of a transaction goes to miner as a transaction fee. The transaction fee depends on the priority of the transaction set by the sender and on the data size of a transaction. In this way, Bitcoin becomes divisible. You can send any fraction of Bitcoin.

For example, in Figure 2-18, I want to pay John 6.9 Bitcoins, I have to use the Bitcoins in three previous transactions, which will have enough Bitcoins to pay John. Therefore, there are three inputs for the current transaction. However, the sum of these three inputs is larger than the amount what I need to pay. I will receive the change back. Therefore, an output of 0.7 Bitcoin is generated as a change for me after paying a transaction fee of 0.1 Bitcoin to the miner. This output is called the Unspent Transaction Output (UTXO). The sum of Bitcoins in the Inputs must be equal to the sum of Bitcoins in the Outputs. The public keys of the inputs become invalid after the transaction.

The transactions shown in the figure are the relevant transactions. For the lightweight nodes, they need only to download the relevant transactions instead of all transactions for the verification. This is called the Simple Payment Verification (SPV). More about SPV will be discussed in Section 3.10.

Miners tend to pick the transactions paying the highest fees to maximize their profit. This means that low-fee-paying transactions could experience extremely long confirmation time or never be confirmed at all. These low fee paying transactions can be floating in the network for a long time. Most likely, the number of these transactions will increase rather than decrease. To solve this problem, Bitcoin Core 0.12.0 introduces opt-in replace-by-fee. If a transaction is sent using opt-in replace-by-fee, users can replace their own transaction with a newer transaction by including a higher fee.[39]

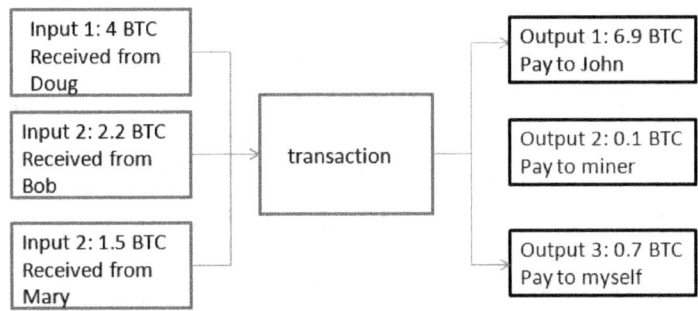

Figure 2-17 Multiple inputs and outputs of a Bitcoin transaction

Since each input came from a different previous transaction, I will have to sign them individually. Imagine that I have received three checks from Doug, Bob, and Mary; I will have to endorse each check separately. However, please note that this is different from the multisig or multi-signature, which refers to the requirement of more than one person to sign the transaction for the security purpose.

Section 2.11 Keep your digital assets safe

When you receive a Bitcoin payment, the sender actually sends the Bitcoin to your public key or your Bitcoin address. Your ownership is guaranteed by your holding of the private key, which unlocks the deposited Bitcoins. Private keys are encrypted with high unpredictability or strength. It is impossible to remember by heart.

Keys are merely a small data file among million files in your disk drives. It is not easy to keep it safe especially if you have many keys. The risk is therefore transferred from the data to the safekeeping of the keys. It is up to you to safeguard your private key(s). If you lose it, you are out of luck. Believe or not, it happens quite often. You can find out a list of "dormant" Bitcoin

---

[39] https://bitcoin.org/en/release/v0.12.0

Bitcoin addresses, which have no movement for at least three years.[40] As of 2016, such dormant Bitcoin addresses held 3.4 million Bitcoins, which has a market value of whopping US$24 billion at Bitcoin price of US$7,000.

Without a third party, you are the only one who is responsible for safeguarding your keys. Lost or stolen keys could compromise the entire assets stored with the keys. It adds more responsibility to the holder of digital assets in the blockchain.

Any backup will create either a risk of theft or risk of the third party if the backup is online. This is a basic dilemma faced by the holders of the digital assets. Deploying a high-assurance crypto management platform is the best way to protect your cryptographic keys.

Without a key manager, maintaining these disparate keys becomes time-consuming and unmanageable. Since keys are being stored in a variety of places, such as USB drives, hard drives, and laptop computers, they are vulnerable to theft and misuse. Backed up keys suffer the same vulnerability.

A crypto management platform can generate, store, and securely manage these keys, so even if data is seized, it cannot be decrypted. Key management centralizes and supports cryptographic keys throughout their lifecycle.

Restricting access to these cryptographic keys is the best practice. The level of security surrounding the key storage container is also important. If the assets in protection are large, one should consider storing their keys in a hardware security module.

A hardware security module (HSM) is a physical computing device that safeguards and manages digital keys for strong authentication and provides crypto processing. These modules traditionally come in the form of a plug-in card or an external device that attaches directly to a computer or network server.

HSMs serve as an isolated, tamper resistant, and responsive platform that acts as a root of trust for the entire

---

[40] https://spreadshare.co/spreadsheet/dormant-bitcoin-addresses

encryption environment. By removing physical key storage devices from networks, you can add another line of defense against security breaches. This is called the Cold Storage, or the off-line storage.

In contrast to the Cold Storage, there is also Hot Storage or online storage. Hot Storage is any storage connected to the network. Therefore, it can be accessed from the network.

The cryptocurrency storage has another name "Wallet". There are many types of wallet where you can keep your private keys instead of just storing as a file in your computer: an app on the mobile device, an app on your computer, hardware wallet, online wallet, or paper wallet. Each type of wallet has its pros and cons.

In each wallet setup, you can have multiple wallets, and you can store many private keys. These multiple wallets allow you to store Bitcoins for different purposes. You can have one wallet for petty cash and one wallet for a large amount of deposit, as the saving account. Of course, you can swap funds between different wallets at any time. When your petty cash wallet balance is low, you can transfer some funds from saving wallet to petty cash wallet by just a few clicks on your computer or mobile device if you have hot storage. If your fund is in a cold storage, you will have to move your fund online. We will discuss each one of these wallets and their advantages and disadvantages.

Web wallet or online wallet is a type of hot storage. Instead of keeping the wallet in your computer, your wallet resides somewhere in the cloud of third parties. The process of accessing Bitcoin is quite easy and convenient. However, the downside is that you have to trust this third party for maintaining the integrity of your wallets, and will allow you to access your Bitcoins whenever you want. They are like banks. However, banks are well established financial institutions but the web wallet services providers are not.

Almost all of the Bitcoin exchanges also functions as wallets. They are like security brokers, where you buy and sell stocks and never bother to get the actual stock certificates from them. This is an advantage for frequent traders. On the other hand,

Bitcoin

you sacrifice your anonymity, since you have to give your identity to open an account with them.

In March 2014, the IRS issued a guidance (IRS Notice 2014-21) labeling cryptocurrency, including Bitcoin, "intangible property." Investors and traders should report the capital gain or loss from cryptocurrency trading, just like stock trading. Web wallets are prone to cyber attack. The most famous examples are MtGox, Bitstamp, and Mintpal. In addition, the network problem at wallet site can prevent you from withdrawing your funds when most needed.

The advantage of web wallet is the ability to bundle transactions together before pushing them onto the blockchain and therefore lowering transaction fees. They can also perform internal transfers for zero fees as an attraction to pull in customers and increase their user base. Some people also find the payment services, often offered by the Bitcoin exchanges, such as Bitpay, are quite useful.

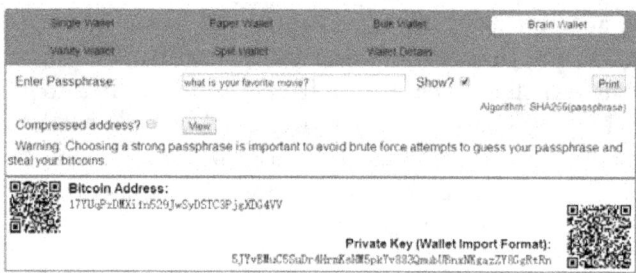

Figure 2-18 A brain wallet

Since the private key is very long and impossible to remember, you can associate a passphrase with your private key. When you go to liteaddress.org or bitaddress.org, it will show you the wallet page as shown in Figure 2-19.

You enter a passphrase of your choice. In this example, the passphrase is "what is your favorite movie?" After you click "View", the program generates the address and private key. Your passphrase can also include spaces and symbols. Like a password, it is also letter case sensitive. Therefore, "What is your favorite movie?" is different from the one used in the example. Every time, if you enter the same passphrase, it will give you the same address and private key.

You can use this address to receive payments and this private key to make payments. As long as you remember the passphrase, you will have the public and private keys. Thus, instead of remembering the private key, you only have to remember the passphrase.

Apparently, this kind of wallet does not have much security. However, it is convenient. It can be used for storing small pocket changes, such as 100 to 200 dollars.

Since the private keys are digital files, they can easily be stored on mobile devices and computers. There are many apps of digital wallets. A typical screen shot of the wallet looks like in Figure 2-20. The wallet can hold multiple digital currencies, and tell you the units you are holding and the amount in USD.

Figure 2-19 A mobile wallet

Another form of web wallet is a hybrid wallet and very similar to the hardware wallet in that the private keys are stored in a different location and under control of the individual, not the third party. Usually, the transaction is sent to the individual to be signed the private key on their computer via JavaScript software written by the provider.

Bitcoin

Desktop wallets are the wallet apps on desktop computers. They work the same way as the wallets on mobile devices. You have to download the app to install the wallet. For example, Bitcoin Core has a wallet function.

In addition to relaying transactions on the network, it also enables you to create a Bitcoin address for sending and receiving the virtual currency and to store the private key for it.

There are many choices of desktop wallets. You can download and try to see which one works better for you. The desktop wallets are in general more powerful. They have more functions. Some wallets are for Bitcoins only. Other wallets can store several cryptocurrencies.

For example, Bitcoin Core Wallet Bitcoin Core is a full Bitcoin client and builds the backbone of the network. It offers high levels of security, privacy, and stability. However, it has fewer features and it takes a lot of space and memory.

Armory Wallet offers backup and encryption features, and it allows secure cold- storage on offline computers. Electrum Wallet emphasizes speed and simplicity, with low resource usage. It uses remote servers that handle the most complicated parts of the Bitcoin platform, and it allows you to recover your wallet from a secret phrase.

To end this discussion, we have to be alerted that the wallets are platform specific. That means, if you have a wallet on your cell phone and a wallet on your laptop, even if you use the same wallet app, say "copay", they are not the same wallet. You will see different balances of your coins in each wallet. The coins you receive in one wallet will not show in the other wallet. It is not a good idea to synchronize the wallets by copying the wallet files, such as "wallet.dat". Many things can go wrong and you will lose your coins. Even changing a directory where your wallet resides can have a certain risk. However, you can share wallet with yourself. For example, you can share your desktop wallet with your cell phone wallet. It will serve the same purpose. Another way to do it is to transfer the coins between your wallets.

If you are fee conscience, some wallets give you full control over fees. This means that this wallet allows changing the fees after funds are sent, or provides fee suggestions based on current network conditions so that your transactions are confirmed in a timely manner without paying more than you have to. As a security measure, most of the wallets also provide 2FA login.

Section 2.12   Two Factor Authentication (2FA)

Two Factor Authentication (2FA) or Multi-Factor Authentication (MFA) is a method that a user is granted access to an object of value (e.g. online account, or a digital wallet) only after the user passes the authentication of two or more factors. When there are only two factors to be authenticated, it is called the Two-factor authentication (2FA). When more than two factors are needed, it is the Multi-factor authentication (MFA).

2FA or MFA is more secure than the simple password or PIN (personal identification number) used to log onto a computer or an online account, or the one-factor authentication. We are using F2A without knowing it in our daily life. When you use the ATM card and PIN to get access to your bank account at an ATM station, you are using 2FA. In such a simple act, you have presented two factors for authentication – the card and PIN. Anybody stealing either one of the factors, the card or the PIN, cannot get access to your account.

There are really three types of factors to prove that you are the claimed owner of the object of value: Something only you know. Something only you have. Moreover, what you are. A password or a PIN is an example of "something only you know". This is the same way we log into most of our online accounts. To guard something more valuable, password/PIN authentication is not sufficient. You will need to show a second authentication factor: Something only you have. Key is a typical example. You open a door with a key – something only you have. In the digital world, this "something" is usually referred to as a token.

A token can be a hardware device issued to the user for the purposes of authentication. For example, a bank ATM card is a token that only you have. Therefore, in order for you to withdraw money from ATM, you need to have both the ATM card and the PIN. This is the 2FA most often used.

Bitcoin

Likewise, hotel issues a card key to you for your room. The card key is something only you have. The key does not have room number and the hotel name. It is the information that only you know.

If this is still not secure enough, then we need to bring a third factor: "what you are". This factor refers to some form of biometric authentication, based on a measurement of some personal characteristics (which may or may not be physical).

Multiple forms of biometric authentication are available today, all of which may enhance the level of confidence in the authentication process. For example, the fingerprint is used in some mobile phones or computers to get access to. Voice recognition and facial recognition are other examples.

The "two-factor" authentication system, which combines something only you know and something only you have, with encryption, is already difficult to crack. There are different kinds of the token (something only you have):

- Paper token: "one time password" is a paper token that only you have, because it is generated at your request (when you sign in an account using password) and send to you only (to your phone) and it is good only for a short period of time, or can only be used once.
- Hardware token: These are physical devices that only the user posses. It is also known as the disconnected token. The hardware token offers the most secure two-factor (token-based) authentication mechanism.
- Soft token: These rely on a "software" component present on the client's computer, e.g. a cookie or a software token application. It is also known as the connected token.

The soft token is the most sophisticated type of token. One example is the Google Authenticator. Google Authenticator is a mobile app developed by Google that implements 2FA service for the Google account.

Google Authenticator uses the Time-based One-time Password Algorithm (TOTP) and/or Hash-based authentication code called Hash-based One-time Password Algorithm (HOTP), for authenticating users of mobile applications. TOTP is an algorithm that computes a one-time password from a shared secret key and the current time. HOTP is one-step further of TOTP. It uses the hash algorithm on TOTP to generate HOTP. Please refer to Figure 2-21 for the flow chart of TOTP and HOTP mechanisms.

To use the Google Authenticator, you have to install Google Authenticator app on your phone. Then follow the instructions to setup two-factor authentications. You have to enter the accounts that you want Google Authenticator to provide you with the temporary password. Once it is set up, each time when you try to log into the account using the password, it will prompt you to enter a one-time-use code generated by the Google Authenticator.

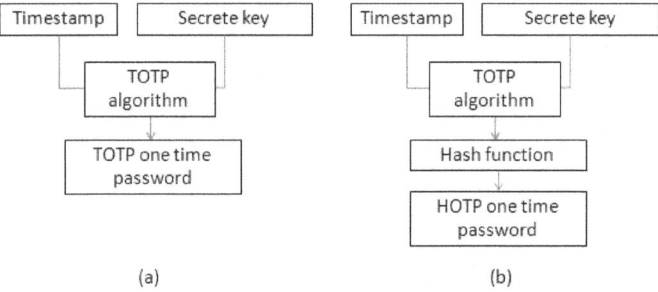

Figure 2-20 Google authenticator

Bitcoin

Chapter 3   Bitcoin issues

Bitcoin platform, although elegantly designed, has some technical limitations and difficulties. Some are fundamental, and some are circumstantial. Over the time, the Bitcoin community resolved many of these issues.[41] However, as the number of Bitcoin transactions grows, new issues are popping up.

Some of these issues are looming so large that there is heated debate as to how to solve them urgently. Without solving them, Bitcoin may run into a bottleneck. Nevertheless, some of the proposed solutions may undermine Bitcoin's fundamental principle of decentralization.

Satoshi Nakamoto designed the Bitcoin with a very lofty goal – no central authority or decentralization. Some of the proposed fixes may undermine Bitcoin's decentralization. To many, it is more important to keep Bitcoin's decentralization. Others believe that without compromise, Bitcoin may find itself less useful to achieve the lofty goal – a cryptocurrency to replace the fiat currency.

One of the problems of the Bitcoin is its block size. Any attempt to increase the block size or attach sidechain to allow faster transactions is bound to make the blockchain more centralized. The most fundamental debate is philosophical: does Bitcoin want to become a store of value or a global payment system? In other words, should Bitcoin behave more like gold or

---

[41] "What is BIP? Why do you need to know about it?", Sudhir Khatwani, https://coinsutra.com/bip-bitcoin-improvement-proposa/

Bitcoin issues

like a currency? After all, people do not gold for daily shopping, and yet it still has its value.

Section 3.01   Bitcoin issues

The Bitcoin popularity has grown tremendously over the last few years. The growing user base increases the number of daily transactions. The number of transactions per day increased tenfold from 34,000 in January 2013 to 350,000 in January 2018.[42] The Bitcoin network is finding itself congested. The congestion has its origin in the software design, which remains the same since it created in 2009.

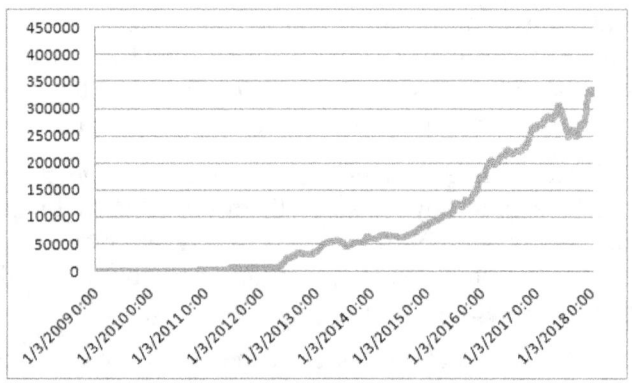

Figure 3-1 Number of Bitcoin transactions per day vs time

New transactions created are broadcasted to the Bitcoin network and received by every miner. The miners compete with each other to validate and confirm the transactions before including them in a new block. Whoever completes the task first, he then attaches his block to the blockchain after other nodes confirm the block.

Every 10 minutes, the mining process creates a new block. The block size is 1 MB by design so that each block can only contain a finite number of transactions. If the new transactions are coming at a faster rate than a block can include them all in a block, there will be a queue of the unconfirmed transactions. The faster the new transactions are coming, the longer is the queue. The

---

[42] https://blockchain.info/charts/n-transactions?timespan=all

queue of unconfirmed transactions resides in the memory pool of the miners' computers. The blockchain.info website continuously updates the size of the queue in memory size.[43] In early 2018, the size is about 100 MB or 100 blocks. Considering the platform generates each block every 10 minutes, 100 blocks will take 1,000 minutes or 17 hours for the miners to digest.

Figure 3-2 shows the average transaction time in minutes[44] from April 2016 to January 2018. The transaction confirmation time increases from less than 20 minutes before May 2016 to more than 400 minutes in June of 2017. The transaction time eased off in August 2017 due to the Bitcoin hardfork, Bitcoin Cash was split from Bitcoin. However, such a relief is only temporarily and the confirmation time shot up to 700 minutes by January 2018.

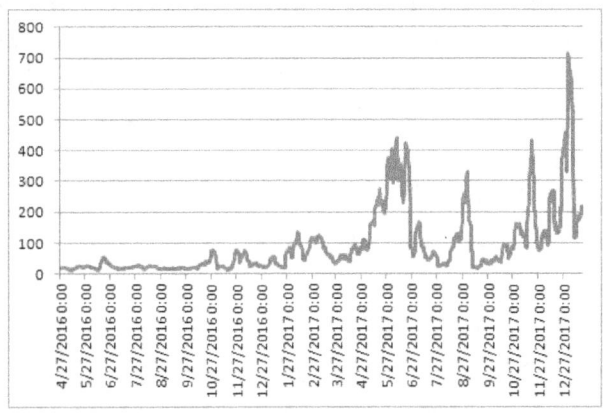

Figure 3-2 Bitcoin transaction confirmation time in minutes from 4/27/2016 to 1/15/2018

The transaction delay is a serious problem because the coins received cannot be spent until the transaction is confirmed. This is the scaling issue due to the limited scale of Bitcoin platform.

Transaction data size varies depending on the amount of data it contains. The transaction involving multiple inputs and

---

[43] https://blockchain.info/charts
[44] https://blockchain.info/charts/avg-confirmation-time

Bitcoin issues

outputs is apparently larger. Figure 3-3 shows the typical distribution of transaction data size.

The transaction data size also tends to increase slightly over time. (Figure 3-4) This is because when the Bitcoin becomes more valuable, most transactions tend to be a fraction of a Bitcoin and therefore involve more inputs and outputs.

The average data size of the transactions is around 400 bytes. 1MB can fit only 2,500 transactions. A block is created every 10 minutes by design. Therefore, the Bitcoin platform can handle only 2,500 transactions every 10 minutes, or 4.17 transactions per second. If the incoming rate of the new transactions is higher than 4.17 per second, the queue of transactions waiting for validation will grow and the transaction confirmation time gets longer. The overflowed transactions remain in the memory pool of the mining nodes and flock to the next block, hopefully, to be included. In early 2017, this is what happened. The Bitcoin blockchain is overwhelmed.

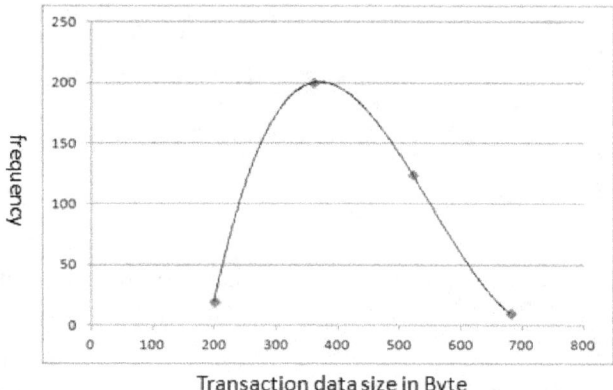

Figure 3-3 Distribution of transaction data size

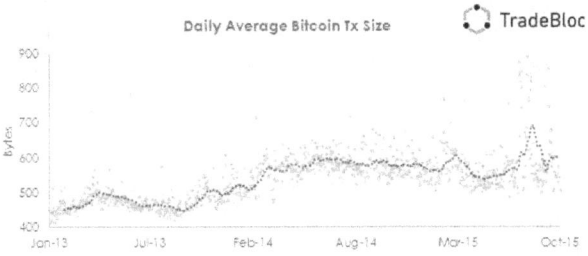

Figure 3-4 Average transaction size

Not only the overflow transactions cause the delay in transaction confirmation time, but also they can crash the system of full nodes if the backlog transactions grow too big to cause the system to run out of the memory. Full node operators can configure their limits to prevent the crash and Bitcoin Core 0.12.0 set the default maximum size to be 300 megabytes.

As a reference, VISA card can handle 24,000 transactions per second or 6,000 times faster than the Bitcoin platform. If Bitcoin platform does not scale up, it will have no chance to become the mainstream transaction platform. It needs to have at least the capability to have transaction speed of VISA today, considering that VISA is only one of the many major credit cards. It seems that the future of Bitcoin, or any other digital currency for that matter, depends on finding a solution to this problem. There are heated debates going on to address this issue.

We will discuss some of the proposed and hotly debated solutions in the next section. However, none of these solutions can improve the transaction rate as to match 24,000 transactions per second of major credit cards.

Blockchain consists of blocks chained together one after another. It records all the history of Bitcoin transactions. The number of blocks in the blockchain continues to grow. In addition, this blockchain is residing on all the nodes of Bitcoin network. Figure 3-4 shows the size of blockchain over time. In January 2018, the blockchain size is around 150 GB.[45] Increasing the block

---

[45] Blockchain size over time, https://blockchain.info/charts/blocks-size

size will improve the transaction rate, but the blockchain size will grow even faster. This will require significantly more resources from the Bitcoin nodes. Unless the cost of computing power can match the block size demand, fewer nodes will be full nodes.

A solution called SegWit, cover in later chapters, aims at reducing the transaction data size by handling signatures off the blockchain.

Section 3.02   Bitcoin and decentralization

The concept of Bitcoin mining is that the mining nodes do the complex work of verifying transactions. They receive newly issued Bitcoins as the reward for their work. In practice, professional miners running huge computing rigs built out of specialized hardware are doing the overwhelming majority of mining nowadays. The larger the hash-rate a single miner control, the more centralized Bitcoin becomes and the more trust using Bitcoin requires.

There are also fundamental trade-offs between scale and decentralization. If the system is too costly people will be forced to trust third parties rather than independently enforcing the system's rules. If the Bitcoin blockchain resource usage, relative to the available technology, is too great, Bitcoin loses its competitive advantages compared to legacy systems because validation will be too costly.

The decentralization factor is important because it is the key differentiator between Cryptocurrency and fiat currency. As such, Bitcoin is more than just a financial instrument, it is also an ideology.

Since its creation in 2008, it has been regarded as a radically different kind of currency that works independently of central banks and the established financial system, with no single centralized source of power or control. Satoshi Nakamoto believes" "As long as a majority of CPU power is controlled by nodes that are not cooperating to attack the network; they'll generate the longest chain and outpace attackers".

As the miners are getting more powerful by amassing huge mining power, the basic assumption of Satoshi Nakamoto becomes less and less valid. By increasing the block size, it effectively weeds out small miners and gives large miners control of the Bitcoin network. Today, the Bitcoin platform is more centralized than a few years ago. When the miners get control of the Bitcoin blockchain, they can do many things in their favor, for example, set higher transaction fees. Eventually, the Bitcoin is no different from today's fiat currencies and miners become bankers. This defeats the intention of Bitcoin.

However, from miners' perspective, the mining fee is an important source of income for them when the Bitcoin reward decreases gradually. Because the lifetime supply of Bitcoin is limited, when reaching the end of Bitcoin mining, there will be no incentive for the miners to do the Proof of Work, without collecting the fees. At that point, no one will verify the transactions, and the Bitcoin platform would collapse. The transaction fee is the only way to keep Bitcoin platform going.

The bitterness of the debate sparked a 20% collapse in Bitcoin price in early 2017, because people thought that the future of Bitcoin was bleak.

While the centralization and decentralization discussion in Bitcoin platform is still largely in favor of decentralization, as we will see in the following discussion that the blockchain technology can be used in both ways and anywhere in between. The irony of the blockchain technology is that when used as Permissioned blockchain, its centralization power can reach an unprecedented level.

The entire decentralization concept lies in the distributed mining power. And the mining power resides in the miners who contribute resources to mine. In the Bitcoin platform, the mining is done using resource intensive PoW process for validation.

As the blockchain technology evolves, less resource intensive validation techniques are being adopted by the altcoins – other cryptocurrencies. It is, therefore, possible that one centralized entity, such as a single company or an organization can afford all the mining resources required, and thus controlling the blockchain. As the Bitcoin blockchain was designed to be decentralized.

Bitcoin issues

A centralized blockchain application can wield tremendous power over its users, more so than any other tools available to mankind today. Such a vision is quickly becoming a reality, as in the case of the program to digitize the Indian currency, rupee, by the Reserve Bank of India,[46] which can track every single rupee spent.

Even Bitcoin, regarded as a truly decentralized system, is built on the premises that the system is secure as long as honest nodes collectively control more CPU power than any cooperating group of attacker nodes. Such premises may not hold.

Section 3.03 MtGox incident

Although Bitcoin is supposed to be un-hackable, its private keys are not. In fact, because of the small file size of the key, it is extremely portable. It can be easily stolen or lost. This is what happened in 2013 when an estimated more than 700,000 Bitcoins in the largest Bitcoin exchange MtGox were stolen since 2011.[47] In Bitcoin's price at US$7,000 at the end of 2017, the stolen Bitcoin is valued at US$5 billion, even though at the time of the theft, the Bitcoin price was around US$480.

The first security breach was discovered in June 2011. MtGox was probably inexperienced in the web security that it did not do enough to patch the problem. Since then, the hacker regularly stole the compromised wallets unnoticed by the exchange. By 2013, MtGox was handling 70% of worldwide Bitcoin transactions. Its trading volume reached 150,000 Bitcoins per day, masking the volume of the stolen Bitcoin. Furthermore,

---

[46] "India's Central Bank Considering Creating Digital Rupee, Dislikes Bitcoin", Lisa Froelings, https://cointelegraph.com/news/indias-central-bank-considering-creating-digital-rupee-dislikes-bitcoin

[47] "The inside story of Mt. Gox Bitcoin's $460 million disaster", Robert McMillan, Wired, https://www.wired.com/2014/03/bitcoin-exchange/

these stolen addresses were being reused, causing confusion in the accounting.

Only when CoinLab, MtGox's partner in the US, filed a lawsuit against MtGox for the breach of contract, thus, calling the attention of the US Department of Homeland Security (DHS), the problem was gradually unearthed. DHS seized the money from MtGox US operation because it was considered as an unregistered money transmitter. As a result, MtGox suspended the US dollar withdraw. Soon afterward, Mizuho bank in Japan closed MtGox's accounts. From that point on, it became extremely difficult for MtGox to operate normally. In February 2014, MtGox had to halt all Bitcoin withdraws and filed bankruptcy protection. Hundreds of millions of dollars worth of Bitcoins went with it.

It is noted that the stolen Bitcoins were from MtGox's hot (online) wallets, while the Bitcoins stored in the cold (offline) wallets were safe. It is a painful but valuable lesson for the theft of Bitcoin or any cryptocurrency for that matter. As the Bitcoin becomes more valuable, and the proliferation of many other cryptocurrencies, the online security risk becomes higher.

Section 3.04　　Full nodes vs. partial nodes

To run a node, the computer needs to download and run a copy of the Bitcoin Core's node software. There are two kinds of nodes in the Bitcoin blockchain network: full node and partial node.

A full node is required to store the complete blockchain, which contains the entire history of the Bitcoin transactions. The size of blockchain is already over 120 GB in 2017. (Figure 3-5) Since the blockchain contains all history of the transactions, it is growing 50 gigabytes per year under the current 1-megabyte block size limit. Besides the memory capacity requirement, the bandwidth needed for the full nodes is also significantly larger, because they need to download and upload all the blocks in the blockchain.

In order to allow more nodes to participate the Bitcoin network, Bitcoin developer eases the resources required. Bitcoin Core 0.12.0 allows nodes to remove older data once their node has verified it. The nodes opt to do are downgraded to partial nodes or lightweight clients status. The partial nodes do not have all the blockchain data. Bitcoin Core 0.12.0 introduces a data cap and a priority for upload traffic.[48] When the data cap is full, the node will skip uploading blocks older than a week.

The lightweight node runs a Simplified Payment Verification (SPV) mode, which only downloads part of the blockchain. They connect to full node clients and use bloom filters so that they only receive transactions necessary and relevant to their operation. Full nodes enforce the consensus rules, but lightweight nodes do not. By doing so, full nodes have consensus power. The full nodes represent the Bitcoin community, which can exercise the economic power. When a fork happens, the branch that has more participation of the full nodes wins.

Each full node stores a blockchain containing only blocks validated by that node. The nodes having the same blocks in their blockchain are in consensus. The validation rules these nodes follow to maintain consensus is the consensus rules. In reality, Bitcoin establishes consensus based on voting among the nodes, which vote with their CPU power, expressing their acceptance of valid blocks by working on extending them and rejecting invalid blocks by refusing to work on them. Since the partial nodes do not have the complete blockchain, they cannot participate in the consensus process. This is to avoid that the partial nodes unknowingly accept a non-valid transaction or block. Thus running the full node is the only way to ensure Bitcoin can provide trust without involving a third party.

---

[48] https://bitcoin.org/en/release/v0.12.0

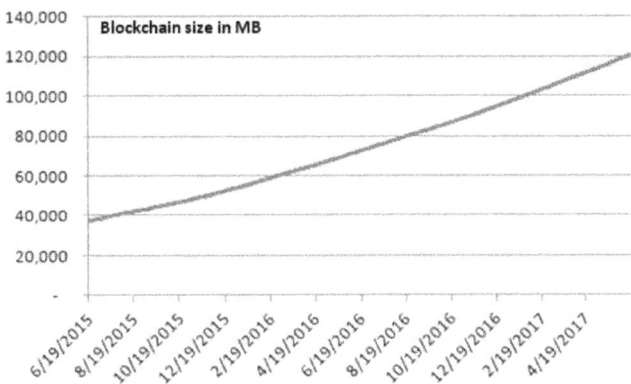

Figure 3-5 Blockchain size in MB

When the blockchain becomes longer, it is more resource intensive. More nodes are opting to run as the partial nodes. This has a consequence that the block accepted by the partial node can be different from that of the full node. This creates a fork. Such a fork propagates until a full node rejects it sooner or later. However, when there are more full nodes in proportion to the partial nodes in the network, there will be more invalid transactions floating around in the network, and the risk increases.

Section 3.05 Is Bitcoin truly anonymous?

Even though Bitcoin network is claimed to provide anonymity, when it connects to the outside world, the user's identity can be revealed in many ways. For example, the identity of Bitcoin users can be revealed from which IP-address a transaction is originated. To protect users' privacy, Bitcoin Core 0.12.0 automatically connects to the Bitcoin network through Tor browser – if the Tor browser is installed on the same computer. Tor encrypts data and routes it through several nodes all across the world before broadcasting it. This makes it hard to trace where a Bitcoin transaction originated.

Bitcoin is best described as pseudo-anonymous in the following way:

Bitcoin issues

- Bitcoin addresses are not tied to the identity of users on a protocol level. When you create a new Bitcoin address, you do not need to submit any personal information.
- Transactions do not link to the identity of users. One can transfer Bitcoin from any address to which it controls the (private) keys to any other address, with no need to reveal any personal information at all. The payee does not need to know the identity of the payer.
- Bitcoin transaction data is transmitted and forwarded by nodes to a random set of nodes on the peer-to-peer network.

When Bitcoin platform interfaces with the real world environment, user's identity can be revealed. Therefore, if one trades Bitcoins in an account linked to his/ her bank account, certainly, he/ she is not anonymous.

Even if you deposit Bitcoins from your cold storage (offline) wallet anonymously, since the Bitcoin platform is transparent and traceable to anyone, someone can trace your identity through one of the many non-anonymous transactions you have made. For example, you use multiple Bitcoin addresses as inputs (please refer to Section 2.10) of a transaction, if one of these inputs can be linked to your identity, all other inputs and outputs can be linked to you.

There are many other analytic techniques, which can deduct the connection between non-anonymous addresses and anonymous addresses, such as taint analysis, amount analysis or timing analysis. [49]

It is not easy to be truly anonymous. If you want more privacy, there are a few simple measures to take: 1) use Tor browser for any Bitcoin transaction to hide your IP address, 2) use a new address for each transaction. Of course, no measure will help your anonymity if your Bitcoin exchange account is linked to your bank account.

Section 3.06 Transaction fee

---

[49] "Is Bitcoin anonymous? A complete beginner's guide", by Aron Van Wirdum, Bitcoin Magazine

Bitcoin miners have incentives to do PoW because of the Bitcoins they receive as reward. However, by design, the reward is halved for every 210,000 blocks mined, while more resources are required for mining due to higher difficulty.

To compensate for the more resource-intensive problem, miners resort to increasing fees for the transaction. We can see in Figure 3-6 that the transaction fee increases as the traffic increases.[50] For Bitcoin transaction of a small dollar amount, the fee charged as Bitcoin amount rather than a percentage becomes excessive. This discourages the use of Bitcoin. The problem is getting worse over time because the mining reward is decreasing and the resource requirement is increasing.

There are factors, which balance the trend of increasing transaction cost. For one thing, the cost of computer power, data storage, and internet speed are improving, and of course, the increasing Bitcoin price also helps.

However, the result of the higher Bitcoin transaction fee is to make Bitcoin uncompetitive. The fee is calculated based on size in bytes in the transaction, not the amount of transaction, which means fees for sending 1 Bitcoin or 0.01 Bitcoin are the same if their transaction size in a byte is the same. The transaction with many inputs and outputs will have a higher fee. It is not reasonable to pay a fee of 0.1 BTC when spending 0.5 BTC.

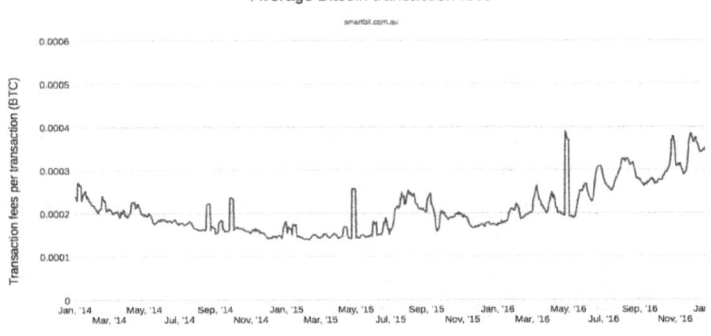

Figure 3-6 Bitcoin transaction fees

---

[50] https://localbitcoins.com/blog/rising-tx-fees/

Bitcoin issues

In most of the wallets, you can set your own fee. When you set higher fee, your transaction will be processed at a higher priority, and your transaction will be confirmed sooner.

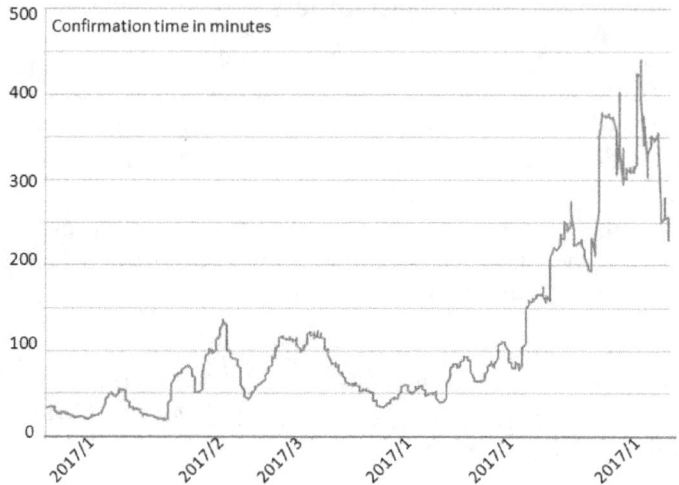

Figure 3-7 Bitcoin transaction confirmation time in minutes

The average time of Bitcoin transaction confirmation is shown in Figure 3-7.[51] The daily average reached an all-time high of almost 500 minutes or roughly 9 hours. But the slower transactions took as long as 2,500 minutes or 2 days to be confirmed during June 2017. This becomes unacceptable.

Section 3.07 Transaction malleability

Malleability means the ability to be shaped. In the Bitcoin world, the transaction malleability means that the transaction can be altered. This can happen to the unconfirmed transactions. The unconfirmed transactions are the transactions broadcasted, but not confirmed. The further transactions created from the Bitcoins of an unconfirmed transaction are also unconfirmed. You will be surprised how many unconfirmed transactions there are at any

---

[51] https://blockchain.info/charts/avg-confirmation-time

given time. [52] At the time of this writing, there are approximately 24,000 unconfirmed transactions in the Bitcoin network at any given time. This number updates itself constantly.

Bitcoin transactions back reference by Transaction ID or TXID, a cryptographic hash of the previous one being spent. The TXID depends on the inputs and signature of the transaction. If the inputs data are changed, the TXID will be different.

In a malleability attack, the attacker can mutate the transaction ID. Such a mutation creates two transactions involving the same Bitcoin. If a miner confirms a mutated transaction before the original transaction, the original transaction will become invalid. Although no one has stolen the Bitcoins, it creates confusion in the Bitcoin network. This also makes it significantly harder to build certain second-layer protocols on top of Bitcoin, like bi-directional payment channels.

Since the exchanges process most of the transactions, malicious software can target exchange for the attack. The attacker picks transactions from the Bitcoin network and modifies signature data creating two different TXIDs circulating on the Bitcoin network for the same transaction. The system has no way of telling the right TXID from wrong TXID. In addition, since a specific transaction can be confirmed only once, just one of the TXID's will be included in a block, while the other will be rejected. Moreover, the transaction with the altered TXID would be included in the block rather than the original TXID.

Since the lower the fee that a transaction pays to the miner, the lower the priority to be confirmed and included in the block, and therefore, the longer is the wait for confirmation. The longer a transaction is unconfirmed; the risk of malleability is higher. Furthermore, if you try to spend the unconfirmed transaction, your outbound transaction will have an "unconfirmed parent". If the parent transaction is tempered, then your outbound transaction will not be able to be validated.

If you a make payments to many parties in a short period, you may split up outputs ahead of time. When the parties receive a 100 BTC output they might break it into four 25 BTC chunks

---

[52] https://blockchain.info/unconfirmed-transactions

Bitcoin issues

paying themselves, which will allow them to pay four people without needing to attempt to spend unconfirmed change. Some delicate balancing is required here if you have too many outputs in your wallet the fees increase dramatically and the confirmation will become slow.

The immediate effect of the malleability attack is that it seems as if the transaction did not go through – though it did with a wrong TXID. In addition, the attack also impedes on chained transactions, even though the attacker cannot steal the fund because only the TXID can be changed, not the transactions themselves.

BIP 62 has a remedy to prevent malleability attacks by narrowing down the types of data that can be included in Bitcoin transactions. BIP 62 has been implemented as a softfork, which means that it is up to miners to adopt the software changes and is backward compatible. [53]

However, BIP 62 does not address all possible malleability issues. Signatures can still be changed by anyone who has access to the corresponding private keys.

Section 3.08 BIP, Hardfork, and Softfork

Bitcoin blockchain is still a relatively young technology. From the early theoretical concept and framework to the real world implementation, many shortcomings are popping up, especially when the scale gets bigger. As the Bitcoin platform evolves, it encounters deficiencies, which requires improvements. Otherwise, Bitcoin platform runs into a bottleneck.

Along the way, there are many proposals to improve the Bitcoin platform. A proposal is different from ideas. It must be implementable. There is a strict procedure for a proposal to become a BIP (Bitcoin Improvement Proposal) by the Bitcoin community. Once it becomes a BIP, it acquires a serial number, and thereafter referred to as BIP {number} by the community. The discussion will start with the potential implementation. Because

---

[53] "How much of the BIP62 has been implemented?", https://bitcoin.stackexchange.com/questions/35904/

Bitcoin platform has no central authority, its implementation or adaption must go through the approval of Bitcoin community consensus.

So far, there are several hundreds of BIP's.[54] Some BIP's are for information only. Others are to change the network protocol, transaction validation, and measures affecting the platform operation. When the change is more dramatic, its implementation is more controversial. Such is the case of SegWit, which deal with the scaling issues.

Any implementation of the changes to the Bitcoin platform will inevitably run into compatibility issue. After all, Bitcoin platform has a set of strict rules. When the rules are changed, there is a new version of the blockchain software that adapts the new rules. However, the blockchain can run only one version of the software.

The blocks formed by the new version will deviate from the existing blockchain. The blockchain will have two heads or a fork. A fork in software development means a spin-off project from the original software project for the improvement of certain aspects of the original software. A fork in the blockchain terminology means two branches of the blockchain grow out of the existing blockchain due to the divergence of software versions used by different miners.

Here, the consensus is at work. It is the fork of the blockchain running the new version against that of the old version. The fork of the blockchain, which wins more hash power or economic power, becomes the majority fork of the blockchain. The other branch of the blockchain is the minority fork. Since the blocks generated by the minority fork will be rejected by the majority fork, the minority fork will either fade away (because no more transactions coming to their way) or join the majority fork by giving up their version of the software. When this happens, the minority fork disappears and the majority fork continues as the only blockchain. The new blockchain can run either the new version or the old version of the software depending on whichever wins the consensus.

---

[54] A list of BIP's. https://github.com/bitcoin/bips

Bitcoin issues

However, if the minority fork persists and does not fade away because there is enough support, then both forks can run in parallel to each other, but each follows a different set of rules. These two parallel blockchains are not compatible and cannot send the transaction to each other because the transaction passing the validation of one fork does not pass the other fork. In another word, the coins are not transferrable across these two blockchains. Two blockchains grow from one. Each blockchain has its own coin.

In some cases, the new rules are backward compatible. The backward compatibility exists when the new rules are a subset of the old rules (Figure 3-8). This type of fork is the softfork. On the other hand, if the old rules are a subset of the new rules, the fork is the hardfork.

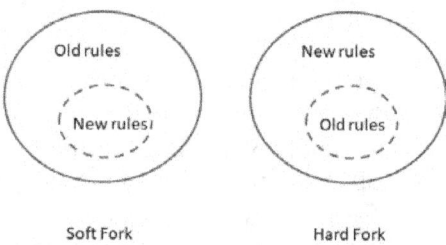

Figure 3-8 In a softfork, new rules are subset of old rules. In a hardfork, the opposite is true.

Figure 3-8 illustrates the difference between two blockchains, one with a softfork and the other with a hardfork. The backward compatibility is outlined in the drawn boundaries. The blocks in the drawn boundaries are backward compatible but not with the blocks outside of the boundary.

In the case of the softfork on the left figure, both Fork A and Fork B are valid by the original blockchain. However, in the case of the hardfork as shown in the right figure, the blocks formed by the new rules (Fork A) are rejected by the original blockchain; therefore, it breaks off from the original blockchain.

Section 3.09   UASF and UAHF

There are two types of consensus mechanisms: by hashing power (computer power) and by economic power. Before 2016, economic power was the only type of consensus. However, as the miner power becomes more concentrated, the weight of hashing power becomes stronger. In 2017, roughly ten miners control the majority of the hash power. [55]

Nodes in a Bitcoin network are not necessary miners, although miners are also nodes. Any computer joining the Bitcoin network is a node. Most of the nodes receive data from other nodes, checking its validity, and passing it on further without doing the mining work. Nodes accept the block only if all transactions in it are valid and not already spent.

The Bitcoin software has a built-in mechanism to reject the transactions that break the consensus rules or the blocks containing such transactions. This prevents the miner to include any invalid transaction. The forks resulted from invalid transactions dissipate quickly and is not an issue. However, if a majority of miners decides not to adopt the change of software mandated by the Bitcoin developer and approved by the Bitcoin community, it will cause a problem. The fork arising because of such a situation is called User Activated.

There are over 10,000 reachable full nodes at any given time. [56] However, the actual number of computers running the full nodes in the Bitcoin network can be much larger. [57] The United States alone has 28% of all the full nodes in the Bitcoin network, followed by Germany 17%, France 6.8%, and China 6.7%. These full nodes constitute the Bitcoin community. Within the Bitcoin community, a small group of full nodes operates as miners. The hashing power resides in the miners, but the economic power resides in the Bitcoin community.

---

[55] "Bitcoin is not ruled by miners", Bitcoin Wiki, https://en.bitcoin.it/wiki/Bitcoin_is_not_ruled_by_miners
[56] "Global Bitcoin Node distribution", Bitnodes, https://en.bitcoin.it/wiki/Clearing_Up_Misconceptions_About_Full_Nodes
[57] "Clearing up misconceptions about full nodes", Bitcoinwiki, https://en.bitcoin.it/wiki/Clearing_Up_Misconceptions_About_Full_Nodes

Bitcoin issues

In the Bitcoin world, the developers hold the key to the software upgrade. They are the ones who develop a new version of software from BIP's. It is up to the Bitcoin community to upgrade the software to the new version by consensus. When the Bitcoin developer imposes a mandatory software version change supported by part of the Bitcoin community, and yet the dissidents are comparable in size, the Bitcoin community splits itself and the User-Activated fork happens.

In a UAHF case, the nodes running the new version will accept the blocks generated by the miners using old rules. The fork of new rules will grow. After the flag date, UAHF will even make valid certain invalid blocks under old rules. It is seen as the voluntary departure of the dissident nodes from the majority blockchain. The old blockchain will continue as if nothing happened. It has minimal impact on the existing blockchain.

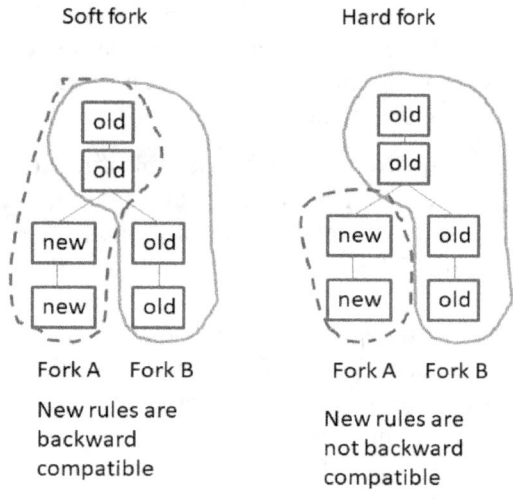

Figure 3-9 The difference between softfork and hardfork

However, for a UASF, the story is very different. The nodes running new rules will not accept the blocks generated by the miners using old rules. Only the miners, who upgrade the software, will be able to generate the blocks for the nodes of new rules. That means, without miners' support, the nodes running new rules will soon be starved of new blocks. This is because the economic power has very little hashing power, if the majority of the hash power does not go along with the new version, there is not enough hash rate to process the transactions. Therefore, even if the economic power wins a UASF, it can be sabotaged by the hashing power.

UASF is like hostile take-over and is also known to as the "51% attack" or "Wipe Out". If someone holds his Bitcoins at a third party, such as exchange, and this third party does not support the UASF, all his Bitcoins can be wiped out or lost. Therefore, it is advised that the owner of the Bitcoins move his coins to his own wallet and keep the private keys before BIP 148 UASF activation.[58]

A UASF, which wins the economic power, still needs miners' support for the attack to be successful. Nevertheless, if the hashing power and economic power are in harmony, UASF can be carried out smoothly. A successful example of the UASF is the implementation of P2SH (BIP16). A fearful UASF example is the BIP 148 activation (SegWit activation).

The most notable hardfork cases are Ethereum forked into the new Ethereum (ETH) and Ethereum classic (ETC) in 2016, and Bitcoin forked into Bitcoin (BTC) and Bitcoin Cash (BCH) in 2017.

Section 3.10   SegWit and change of block size

The Bitcoin issues related to the transaction congestion are generally known as the Bitcoin scaling because they are caused by the limited Block size of one megabyte in the original software design. Each transaction is about 400 bytes big; the number of transactions per block is limited to around 2,500 by 1 MB block size. Since only one new block is generated every 10 minutes, Bitcoin cannot handle more than 250 transactions per minutes, on average. To solve the scaling problem is to break this barrier. There are two approaches: One is to make the transaction data smaller, the other is to increase block size.

---

[58] "How To Survive BIP148 The Upcoming Bitcoin UASF", Bitcoin Chaser, http://bitcoinchaser.com/survive-bip-148

Bitcoin issues

Each transaction contains 6 to 8 inputs and outputs. Roughly, 60% of the data consists of signature(s), a part of the input, which verifies that the sender has the funds to make a payment.

The most prominent proposal to solve this problem is called SegWit, short for Segregated Witness. It is a proposed software update by Bitcoin Core to fix the Bitcoin scaling and the malleability issues in December 2015. The Bitcoin Core software, version 0.13.1, was released in October 2016.[59] It includes the SegWit fix and other minor changes, such as the increased security for multisig. In order to avoid a contentious fork, the developers of SegWit sets a rule for activation with at least 95% vote.

SegWit proposes to separate the digital signature from the transaction data by stripping off the signature from the input and move it to a structure towards the end of the transaction. This reduces the transaction data size.

SegWit adds another layer to the blockchain. Now the signature is handled in a different layer from the transactions. The additional layer is called the Sidechain. SegWit is also intended to solve the malleability problem. Since the digital signature would be detached from the input, the attacker would have no way of changing the transaction ID without also nullifying the digital signature. The proposal also includes fraud proofs and Simplified Payment Verification or SPV.

Nevertheless, the upside of scaling is limited. After all, even if SegWit can reduce the transaction size by half, the number of transactions that 1MB block can contain is only doubled. It is light years away from the transaction capacity of any credit card. Therefore, many people think that SegWit is only a temporary fix.

In May 2017, miners that represent 80% of the Bitcoin hashing power got together in New York and signed a so-called the New York agreement. In this agreement, in addition to the SegWit implementation, they also proposed to double the size of block size to 2 MB in six months following the implementation of

---

[59] Bitcoin Core 0.13.1 released. https://bitcoin.org/en/release/v0.13.1

SegWit. This effort is dubbed as SegWit 2X. With SegWit 2X, the block size is doubled again. But that is still not nearly enough.

Many people are against SegWit because potentially it can degrade decentralization. In the second-layer sidechain implemented by SegWit solution, a trusted third party, without the need to broadcast them across the entire network, processes all transactions. Over the long run, the third party running the sidechain will become the centralized authority of the blockchain. It defeats the original purpose of the blockchain to be the completely decentralized system.

In summary, the SegWit solution does not solve the long-term scalability issue, while it will potentially undermine the decentralization of the Bitcoin network. Nevertheless, there is a pressing issue of congestion needed to be solved, and SegWit can bring immediate relief probably for one or two years, which buys time for the community to find a permanent solution. The Segwit softfork was implemented in August 2017, and the increase of block size to 2MB was targeted for November 2017 implementation. But it did not happen due to the lack of support.

Section 3.11   Bitcoin split

After ongoing debates over how to scale Bitcoin, Bitcoin was officially split into Bitcoin and Bitcoin Cash after a hardfork on August 1st, 2017. Bitcoin Cash (BCH) started trading on August, 3rd, 2017. It is the first Bitcoin hardfork. For the owner of a Bitcoin before the hardfork, he now got two coins: Bitcoin and Bitcoin Cash.

Since the dissident group, mostly miners and some developers initiated the hardfork, it is also known as the User-Activated Hardfork. Upon the UAHF, a copy of Bitcoin's blockchain from block 478,558 was transferred to the BCH blockchain.[60]

---

[60] "Bitcoin vs. Bitcoin Cash: What is the difference?", Jake Frankenfield, https://www.investopedia.com/tech/bitcoin-vs-bitcoin-cash-whats-difference/

Bitcoin issues

After the Bitcoin split, SegWit was activated in the Bitcoin blockchain and subsequently gains significant support from the mining industry.[61]

Bitcoin Cash did not implement SegWit. Instead, the block size limit became adjustable, with an initial increase to 8MB. It can be further increased depending on the future need.[62] In fact, the Bitcoin Cash developer makes it easier for the future change of the block size by using a full node implementation of the Bitcoin Cash protocol, called Bitcoin ABC.[63] ABC means Adjustable Bitcoin Cap.

With a future roadmap of massive scaling, Bitcoin ABC allows an immediate block size increase with a simple, sensible, adjustable block size cap. Therefore, the scaling problem has been resolved once for all. It paved the way for Bitcoin cash as cryptocurrency for payment.

There are other changes as well, such as the new transaction type using SigHash type signature, which provides replay, and wipeout protection improves hardware wallet security and eliminates the quadratic hashing problem. This essentially minimizes user disruption and permits safe and peaceful coexistence of the two chains. There is a new PoW Difficulty Adjustment Algorithm (DAA), which allows miners to migrate from the legacy Bitcoin chain as desired.

Bitcoin Cash has attracted support from users who want block size increase, as well as developers of other proposals such as Bitcoin Classic and Bitcoin Unlimited. Most important of all, Bitcoin Cash has the support of some big miners, such as ViaBTC, which has 4% of the Bitcoin hashing power, Bitmain's Antpool, BTC.top, Viabtc, and Connect BTC, etc. Many think that if Bitcoin fails to implement the 2x part of the SegWit 2X, as it happened on November 16, 2017, the scheduled date for 2X implementation, many miners will likely defect to Bitcoin Cash.

---

[61] "Segregated Witness Activates on Bitcoin: This is What to Expect", Bitcoin Magazine, https://bitcoinmagazine.com/articles/segregated-witness-activates-bitcoin-what-expect/
[62] https://www.bitcoincash.org/
[63] https://www.bitcoinabc.org/

There are also mining companies, like Bitmain, which supports both Segwit2x and Bitcoin Cash. Like Bitcoin, no single entity controls Bitcoin Cash. Only coordination across independent development teams can make changes.

Bitcoin price was $2,754 just before split on August 1, 2017. By December 2017, the Bitcoin price is well over $9,700 and Bitcoin Cash is over $1,700. Since each Bitcoin split into a Bitcoin and Bitcoin Cash, the value of Bitcoin has quadrupled since the split in a short 4 months. This is probably the best investment ever.

The Bitcoin split created two very different cryptocurrencies: one for payment and transaction, one for the store of value. (Figure 3-10)

If Bitcoin behaves like a store of value, many financial derivative products will also be applicable to the Bitcoin. Some financial institutions start to launch Bitcoin in the future market. On December 1, 2017, The U.S. Commodity Futures Trading Commission (CFCT) approved Bitcoin for the listing of such products from CBOE, CME Group, and Cantor Fitzgerald.[64] JPMorgan Chase & Co. is also planning to offer Bitcoin futures contract through its CME futures exchange.[65] Whether this product, XBT, will take off, it's anybody's guess. Bitcoin price is already volatile enough; its future contract will be even more volatile.

After all, if gold can be traded in the future market, why not Bitcoin. In fact, the Old Mutual Gold and Silver Fund is the first traditional precious metal mutual fund to invest in Bitcoin.[66]

---

[64] "Bitcoin price above $10,500 on US future listings", Omkar Godbole, Coindesk, https://www.coindesk.com/bitcoin-price-pushes-above-10500-on-u-s-futures-listings/

[65] Alexander Osipovich, "Maybe Bitcoin isn't untouchable at J. P. Morgan after all", Wall Street Journal, Nov., 21, 2017

[66] "A gold fund is investing in Bitcoin", Ryan Browne, CNBC, https://www.cnbc.com/2017/11/24/a-gold-fund-is-investing-in-bitcoin.html

Bitcoin issues

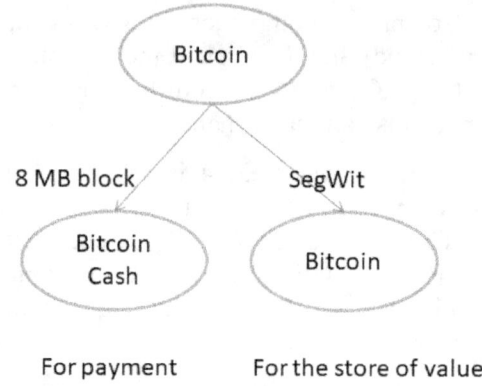

Figure 3-10 Bitcoin split

On the other hand, it can be an interesting experiment, which may evolve into something else. In fact, CME Group already has blockchain project for the gold trading.

Section 3.12 Keep your coins safe during forking

In the future, forking will be a more frequent event. It is important to know the implication and risk of Bitcoin or any other coin split. The replay attacks and the loss of coin value are the only two primary ways in which a potential network split could affect the coin holders. There are several contingency measures to secure your wealth during the period preceding a potential hardfork and immediately after it.

First, the replay attacks. It is a form of attack on a freshly split network when the two blockchains are barely different from each other. When a person sends his or her Bitcoins to someone, an attacker may replicate the same transaction on the other Blockchain and the network will accept it. Thus, unknowingly and unwillingly, the user may lose their coins on the alternative Blockchain just by sending transactions across the first one.

In order to prevent this, it is advisable to refrain from conducting any transactions at all, until the dust is settled and normal transaction volume returns. Of course, the new version of the software causing forking does also implement preventive measures against a replay attack.

Another important action of safety is always to be in control of the private keys to your funds - which may mean withdrawing your coins from an online exchange or wallet to a software wallet stored on your PC.

One of the very purposes of Bitcoin is to give its users complete reign over their coins. By delegating that control to any third party, even one you deem trustworthy, such as an exchange, or a web wallet service, you subject your coins to unnecessary risk. The risk is even more unwarranted during such a turbulent time as a contentious hardfork.

When users are not in control of the key to their funds, they are at the mercy of the decisions made by the company holding the keys.

Hardfork split is like a stock split. The moment a hardfork takes place, all the people holding Bitcoins receive the equivalent amount of coins on the alternative Blockchain. That is because all the records on both Blockchains are identical up until the point of the fork. Being left with two sets of coins, some people will want to dump the ones they find less promising. That may provoke more people into panicking and also trying to sell before the price drops too much. All that commotion creates a lot of uncertainty and volatility on the market, which is bound to last for some time after a hardfork takes place.

For a short while after the split, the combined value of two new coins may be less than the value of the old coin. Under such conditions, the least risky strategy is to just hold onto your coins on both Blockchains and wait it out. It is impossible to predict which one of the alternatives will be the most successful, and taking sides may result in a massive loss if you choose wrong.

A much safer solution is to wait for the market to calm down, see which coin comes out on top and go from there. In other words, if you truly believe in the eventual success of Bitcoin, your best bet is to keep holding. No matter how dire the short-term consequences of a fork can be, the community will eventually sort it out, like it always did.

Bitcoin issues

Section 3.13   Other proposed fixes

Beyond SegWit, there are many other interesting but less visible proposals, such as Invertible Bloom Lookup Tables (IBLT) scheme,[67] P2Pool,[68] Weak block,[69] and others so-called "non-bandwidth" scaling schemes.

There are many the so-called non-bandwidth solutions to fix malleability issues. For example, there are proposals of the bidirectional payment channel and transaction cut-through. There are also proposals for a flexible block size that allows miners to produce larger blocks at some cost.[70] These proposals preserve the alignment of incentives between miners and general node operators and prevent defection between the miners from undermining the fee charging market behavior.

Bitcoin Unlimited is the alternative software client for the Bitcoin network. It is a big proponent of the flexible block size. Bitcoin Unlimited was the first client to provide the tools that node operators and miners need to remove the 1 MB block size limit. Its software allows users to find the limit having a majority consensus and set their block size limit to that value. The node operators running BU client can easily adjust the size of blocks their node accepts, without having to restart their node or recompile new executables from source code.

Giving miners the option to configure the size of blocks they will validate, Bitcoin Unlimited is doing things very differently from Bitcoin Core. The default size is still one megabyte, yet the maximum generation size is a fully customizable option.

---

[67] "Bitcoin in Bloom: How IBLT allows Bitcoin to scale", Alex Gorale, https://www.crypto coinsnews.com/bitcoin-in-bloom-how-iblts-allow-bitcoin-scale/

[68] https://en.bitcoin.it/wiki/P2Pool

[69] "IBLT and Weak Block propagation Performance", https://scalingbitcoin.org/hongkong2015/presentations/DAY1/3_block_propagation_1_rosenbaum.pdf

[70] "A flexible limit: trading subsidy for larger blocks", Mark Friedenbach, https://scalingbitcoin.org/hongkong2015/presentations /DAY2/3_tweaking_the_chain_2_friedenbach.pdf

Network nodes have a similar option to determine which blocks they will validate and how large they can be. As one would expect, Bitcoin Unlimited has gained some support from mining pools. However, the mining community is only a part of the overall network. Service providers, exchanges, and wallet operators are not too keen on what Bitcoin Unlimited proposes. Without that critical support, the chances of BU succeeding are limited.

Another interesting solution to the Bitcoin scaling problem is the micropayment. Micropayment provides a radically different solution than the SegWit or larger block size to achieve the transaction rate faster than the credit card transaction rate at over 30,000 per second by moving the payment mechanism outside of blockchain.

Micropayment is to set up a pending transfer by dividing a large payment amount into many small payments. This is like that instead of sending a $100 bill for a payment; you send 100 $1 dollar bills in 100 separate envelopes. The payment channel is off the blockchain so that the transaction data in the blockchain is minimized. In this way, it reliefs the congestion of transaction traffic in the blockchain.

Currently, such a micropayment method already exists on the Bitcoin platform. Satoshi Nakamoto himself disclosed it in an email on January 1, 2009. Only that it is implemented by offloading the transactions to a third party custodian, which creates counterparty risk. The current micropayment channels in the Bitcoin platform still use real Bitcoin transactions, only electing to defer the broadcast to the blockchain in order for both parties to guarantee their current balance on the blockchain. Therefore, it is slow and expensive. However, there are other micropayment solutions in development, as we will discuss in the next section.

Section 3.14 Lightning network

Bitcoin issues

The Lightning Network[71] is a micropayments-enabling low cost and instantaneous network that operates bi-directional off-blockchain payment channels, sometimes called microchannels. Channels can be created by funding transactions. The funding amount is the maximum spending you can have on this channel. The payment is sent in the form of redeemable IOU's. It also works like a debit card. It promises to be able to process thousands of transactions per second instead of every 10 minutes as Bitcoin today, and best of all, there is no transaction fee.

Joseph Poon and Thaddeus Dryja first proposed the idea in their white paper. Multiple developers developed the software. The framework of Lightning Network is set up to encourage multiple developers to contribute to the development. Functionalities can be introduced without affecting all Bitcoin users. According to its developers, Lightning Network is safer and more reliably than SegWit.[72] Bitfury Group[73], a full-service blockchain technology company, has tested the Lightning Network algorithm successfully.

The Lightning Network uses the built-in blockchain smart contracts with the same security and immutability of the Bitcoin blockchain itself. The solution is to overcome the delay in the Bitcoin micropayment transaction and to avoid the counterparty risk.

In the Lightning Network, parties use multisig addresses between them for a predetermined period, or the nLockTime for the transaction, which creates a ledger entry on the blockchain. Both parties need to fund the ledger in order to open the channel.

They also create the refund transactions, for refunding them in case the transaction fails (say, one party does not honor the contract). However, this will not happen until the nLockTime elapses to allow sufficient time to complete the transaction.

---

[71] "The Bitcoin lightning network: scalable off-chain instant payments", Joseph Poon, Thaddeus Dryja, The lightning network white paper.

[72] "Understanding the Lightning Network", Aaron van Wirdum, Bitcoin Magazine

[73] http://bitfury.com/

They can transact as many times as they want between them as long as the channel is open (without broadcasting the transaction to the blockchain). The ledger updates itself instantly after each new transaction. When they decide to close the channel, either one of them can broadcast the transaction to the blockchain. Upon closing the channel, the parties receive refunds of the initial funding.

This is somewhat like when you pay the cashier by coins; you lay out the coins on the table one by one and count to the cashiers. The cashier will not take any coin until both sides confirm that the sum of the total coins is the transaction amount. Only then, the cashier takes the coins and rings the cash register.

Once the transaction is broadcasted to the Bitcoin network and the channel is closed, the transaction proceeds normally like any other Bitcoin transaction. The multi-micropayments will appear as a single transaction to the Bitcoin platform. This reduces the loading of the Bitcoin transaction and effectively increases the block capacity without touching the blockchain functions.

Since there is no limit on how many channels can be opened in the Bitcoin network, Bitcoin scalability can be achieved using a large network of micropayment channels. This implementation does not require a soft-fork and may be used on the Bitcoin blockchain today with some minimal risk.

Not only the channel can be bi-directional, but also involves more than two parties. When there is no direct channel between two parties, they can create a transaction path through the existing open channels, involving third parties. Figure 3-11 shows the payment from Alice to Bob through the four existing channels (Alice to A, A to B, B to C, C to Bob) in the Lightning Network. [74]

The ledger entry data are sent over the network similar to routing packets on the internet. The nodes along the path need not be trustworthy, as the payment is processed by a script in the smart contract, which enforces the pass-and-fail of the entire payment

---

[74] LND overview and development guide, Lightning Network Developers, http://dev.lightning.community/overview/#payment-channels

through a time-lock. Anyone intercepts the micropayment will not be able to use it since they are only pockets of information, which will not turn into a Bitcoin transaction until the transaction is processed by the Bitcoin network. In this way, the transaction is secure without a third party. By doing so, transactions can be done off-blockchain and with the confidence of the blockchain enforceability.

In order to enhance the trust when the third party is involved, a Hashed Time Lock Contract (HTLC) is used. Since many transactions are off-blockchain, malleability poses a bigger problem. Therefore, the implementation of Lightning Network solution is possible only after SegWit activated in the blockchain.

Although miners do not receive fees for the off-blockchain transactions, the miners receive the benefit of dramatically increased transaction volume. The increased transaction volume will bring more fees as income to the miners.

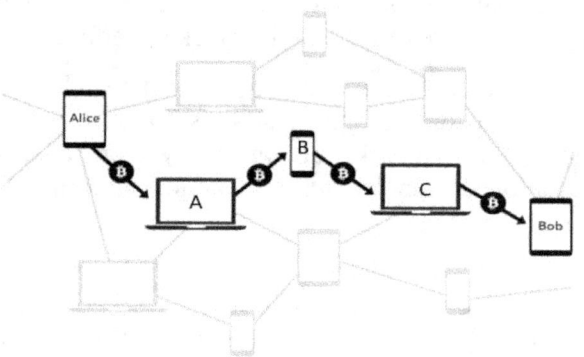

Figure 3-11 Payment from Alice to Bob through third

With the activation of SegWit in the Bitcoin chain in August 2017, Lightning Network is one-step closer to the implementation. The Lightning Lab has released a new version called Lightning Network Daemon (LND) in the same month. The new version added a Litecoin operating mode for the cross chain atomic swap, which meets the requirements of all the swapping parties. LDN can manage channels on both Bitcoin and Litecoin blockchains, and allow transfer between these two chains. [75]

The lightning solution has the potential to Stream Money or allowing instant Payment for Value. The ideal of Payment for Value or Value-Bases-Payment (VBP) came from the healthcare industry. Value-Based Payment is a strategy used by purchasers to promote quality and value of healthcare services. The goal of VBP program is to shift from pure volume-based payment, as exemplified by fee-for-service payments to payments more closely related to outcomes. Such a strategy came from the earlier Payment-for-Performance (P4P) strategy.

The Instant Payment for Value can apply beyond the healthcare industry, it will be an essential development for IoT where machines will generate billions of transactions per second and clear the transactions immediately.

A structural shift in commerce that creates market liquidity can transform the retail industry by leveling the playing field for smaller businesses to get paid online. Lightning Network can unlock Bitcoins massive potential. For example, its capability of handling micropayments using the eight decimals places available allows people to trade tiny amounts of value, where on the current Bitcoin payment rails is uneconomic to do so.

However, Lightning Network is not without criticisms. [76] [77] Some think the Lightning Network can lead to centralization. After all, the banks have all the resources to provide thousands of open channels.

Section 3.15   Government attitude

---

[75] "What is the Lightning Network Daemon", JP Buntinx, http://themerkle.com

[76] "Mathematical proof that Lightning Network cannot be a decentralized Bitcoin solution", Jonald Fyoodball, http://medium.com

[77] "Simulating a decentralized network with 10 million users", Diane Reynolds, http://hackernoon.com

Bitcoin issues

Governments around the world look at Bitcoin with suspicion because of its anonymous nature and free flow across the national boundaries.

As of October 2013, many countries have completely banned Bitcoin: Bangladesh, Bolivia, Ecuador, Iceland, Kyrgyzstan, Russia, Vietnam, and others. Other countries set strict rules to deal with Bitcoins. China banned financial institutions from dealing in the virtual currency as of December 2013, although trading volume in Chinese Yuan persists. Germany, France, Korea, and Thailand have all looked unfavorably on Bitcoin. The European Banking Authority, Switzerland, Poland, Canada, and the United States continue to deliberate about different Bitcoin-related issues.

Countries try to match up Bitcoin to their existing regulatory structures, often finding that cryptocurrencies do not quite fit and ultimately concluding that new legislation is required. However, new legislation takes time. Especially the blockchain technology is still evolving fast, there is no telling how it may become in a few years. Meanwhile, blockchain technology is branching out as Fintech, which takes a whole new life on its own. As more financial industries are embracing Fintech, hasty legislation may become obsolete as soon as it becomes effective.

At present, some countries, like the UK, have classified Bitcoin as a currency, whereas other countries treat Bitcoin as an investment, in the same category as stocks and bonds. While some cryptocurrencies may become currency, others may become the store of value, just like gold.

Even different agencies in the same country may treat cryptocurrency different. For example, In the US, the Internal Revenue Service treats Bitcoin like stock, that Bitcoin is subject to capital gains taxes on transactions. IRA requested cryptocurrency exchanges to report annual transactions over $20,000.[78] However, there is other US government agency—including FinCEN (financial crimes enforcement network), banking regulators, and the CFPB, SEC, CFTC, and DOJ—regulate Bitcoin as a currency.

---

[78] "IRS blinks in Bitcoin probe, exempts Coinbase transactions under $20,000", Fortune Magazine, http://fortune.com/2017/07/10 /bitcoin-irs-coinbase/

Due to the centralized nature of the Permissioned Blockchain, many countries also either contemplate or plan to issue digital currency. With internet access widely available and encryption technology improve, the conditions are ripe for digital currencies, which can reduce operating costs, increase efficiency and enable a wide range of new applications. China's central bank, the People Bank of China (PBOC), express the desire to supervise private digital currencies and develop its own national digital money. Its intention is to maintain financial stability, innovation, and proper supervision on the issuance and circulation of its legal digital tender. Digital currency can co-exist with cash for quite a long time before it eventually replaces cash.

In fact, most central banks in the world are experimenting central bank issued digital currency in different stages. They include Canada, Denmark, UK, Sweden, Germany, Netherland, Russia, Ukraine, India, Ecuador, South Korea to name just a few. Among them, India probably has the most advanced and aggressive program for its digital currency. We will discuss in more detail in Chapter 6.

As far as Bitcoin is concerned, despite the government crackdown, such as in the case of China, it cannot damp Bitcoin for long. Each time the government takes a severe measure to outlaw Bitcoin, the Bitcoin price is inevitably depressed, but not for long. When it bounced back, it did so with much vigor. For example, in early September 2017, China banned Bitcoin and drove the price of Bitcoin down from $4,500 to $3,700. However, one month later, the price of Bitcoin bounced back to $4,500, and since then, it went as high as $11,000 In November. The action taken in China has a big impact because China mines 80% of the worldwide Bitcoin.[79] Even so, it can do the minimum to dampen the momentum of Bitcoin.

---

[79] "The electricity to mine Bitcoin this year is bigger than the annual usage of 159 countries", Oscar-Williams Gruts, Business Insider, http://www.businessinsider.com/bitcoin-mining-electricity-usage-2017-11

Chapter 4   Consensus mechanisms

The consensus in a network is the process of reaching an agreement. Reaching or achieving the consensus means the majority of the nodes agree on something or some event (e.g. that a block or a transaction is valid). Once they reach a decision, that decision is final. Simply put, the consensus is a form of agreement in the distributed system.

There are different types of consensus protocols depending on the network setting. There are in general three types of network setting from consensus perspective- public, private and federated. The meaning of public and private network is obvious. The federated network or a federation is a group of distinct, disconnected networks, usually private agreeing on a standard of operations so that user from one network can send a message to the user in the other network. One example is the Yahoo messenger and MSN messenger.

This concept applies to databases as well. A federated database system consists of several independent databases that can allow users to store and retrieve data from multiple noncontiguous databases with a single query. This is possible through a common standard, such as user interface. Obviously, the consensus protocols for the public, private and federated networks are very different. Even within the same type of network, there are many consensus protocols.

Bitcoin, a public blockchain, uses Proof of Work as consensus mechanism for validation. PoW is a resource-intensive mechanism. According to some estimates, the Bitcoin network consumes power at an annual rate of 32TWh—about as much as

Consensus mechanisms

Denmark. [80,81] This is clearly not sustainable. In fact, the best choice of using PoW as the Bitcoin consensus mechanism may not be a good choice for blockchain of other applications.

A blockchain platform may use other generic consensus mechanism and implement it with the protocol matching its trust model. The choice of consensus mechanism will not affect blockchain properties like distribution, cryptographic immutability, and transparency. Indeed, there are many other consensus/ validation mechanisms available, which are equally effective and less energy intensive. They have received attention for other blockchain systems.

In a permissionless blockchain, transaction validation and establishing consensus is inherently together such as in Bitcoin and Ethereum. In a permissioned blockchain protocol, one could separate the consensus and the validation of transactions.

Building consensus protocol belongs to a discipline of computer science. Consensus protocols evolve from cryptosystems and other security mechanisms. Their assumptions, security models, logic/ reasoning must endure serious challenges. Developing consensus protocols is similar to engineering cryptographic systems, which requires experience in cryptography, security, and the theory of distributed systems. All aspects of their performance and their resilience to actual attacks or network failures are important.

Building a consensus protocol is a difficult task. Among the recent flurry of blockchain-consensus protocols, many have not progressed past the stage of a paper-based description. A consensus protocol must perform under a wide range of adversarial conditions on the nodes and the network, including malicious attacks.

There are many possible consensus models. In fact, no single consensus model can fit all the applications. Some most

---

[80] "Bitcoin's insane energy consumption, explained", Timothy Lee, https://arstechnica.com/tech-policy/2017/12/bitcoins-insane-energy-consumption-explained/

[81] "Bitcoin energy consumption index", https://digiconomist.net/bitcoin-energy-consumption

discussed consensus models are PoW, PoS, Synchronous or Asynchronous Byzantine Agreement, Byzantine Altruistic Rational Fault Tolerance, Federated Byzantine Agreement, and many others.

Different cryptocurrencies use different consensus protocols. It is safe to say that the fate of these cryptocurrencies depends largely on the success of their consensus protocol. A protocol may look great on paper, but when put into practice, it does not perform as expected. Detailed analysis and formal argumentation are necessary to gain confidence that a protocol achieves its goal. In fact, many cryptocurrencies may fail because their consensus mechanism may not endure the real world challenges.

When the blockchain applications are more private-oriented, traditional trust mechanism can offer an alternative solution that sometimes is more preferred than the consensus mechanism for the ease of integration of blockchain applications into the existing financial system. A combination of the traditional trust mechanism and the cryptographic consensus mechanism can offer advantages of both worlds.

Section 4.01 Proof of Work vs. Proof of Stake

Bitcoin blockchain uses Proof of Work (PoW) as the consensus, which is an elaborative process that the miners do to validate the transactions. It requires huge computing power, thus the electricity resources and is ecologically unfriendly. It is one of the major downsides of Bitcoin mining, but it provides security to the Bitcoin system. Because PoW requires large resources, an attacker also needs to have comparable resources to launch an attack. In this sense, any method to reduce the resources is not compatible with the security using the basic PoW methodology.

To overcome the energy consumption problem in PoW, some modified PoW systems allow transactions to include a hash of a recent block known to a transaction sender. The nodes will reject the transaction without the hash of reference block.

Consensus mechanisms

A popular alternative consensus mechanism is the Proof of Stake (PoS). In PoS, user's ownership stake in the system is used as a security deposit in the validating process instead of using the brutal force of computing power to solve the puzzle in the PoW mining process. The rationale is that one is unlikely to attack something that he himself is the biggest stakeholder. [82]

PoW and PoS are two opposite mechanisms on validating the blocks. Block mining is the process of solving a computational challenge imposed by a PoW protocol. On the other hand, block minting is the process of collateralized validation in a PoS protocol. The word "minting" implies that you need to have currency reserve as a backup to mint your coins. It is the same concept that some central banks have a gold reserve to back up their printing of the national currency. In the PoS system, the minter put an ownership stake in the system as a security deposit.

Sometimes, PoS is also called virtual mining. The PoS miners do not receive new coins as a reward as an incentive, but rather receive transaction fees. [83] They are virtual miners or validators. Their stake of not following the rules is that they can lose their deposits. The downside of PoS implementation is that it is more vulnerable to attacks than PoW.

The PoS consensus is not as objective as PoW consensus. In an objective consensus protocol, a new node equipped with only the knowledge of protocol definitions must arrive at the same state as the rest of the network. While this is true in the PoW consensus, it is not true in the PoS consensus. This is because, in PoW consensus, the new node will always regard the blockchain with the highest computational difficulty as the valid chain. Forks pose no threat to PoW because they cannot pass the consensus rules.

In the PoS system, it is possible that two or more nodes can come to two different but equally valid conclusions due to the asynchronous nature of the network that the arrival order of

---

[82] "Proof of Stake", Bitcoin Wiki,
https://en.bitcoin.it/wiki/Proof_of_Stake
[83] "PoW vs. PoS: Basic mining guide",
https://blockgeeks.com/guides/proof-of-work-vs-proof-of-stake/

transactions is at varying times among nodes.[84] In addition, because of the low resources required, miners can mine multiple blocks simultaneously to make sure that one of the blocks will win. No matter which fork prevails, the miner always gets the reward without any penalty. This is the "Nothing at Stake" problem. In this sense, the PoS system is more chaotic.

In addition, the PoS genesis block members can always launch a fork with an alternate transaction history and take over the original fork, no matter how long the blockchain is. Therefore, the PoS scheme weakens decentralization and mathematical soundness. In order to prevent long-range forks from occurring, a PoS system needs to implement subjectivity by combining protocol rules with the social-driven security.

However, some may argue that in the PoW scheme, as the mining process becomes more resource intensive, the nodes with high hashing power will always dominate. Eventually, a few miners with high hashing power will mine all the blocks. This is a form of centralization in a decentralized system. Therefore, PoW is not perfect either.

PoS is a viable alternative to PoW for certain applications, although it has not been proven as trustworthy in the field as PoW. PoS cryptocurrencies are taking precautionary measures to prevent the "Nothing at Stake" vulnerability. A recent development in PoS is the so-called Delegated PoS system or DPoS.[85] It provides a solution to the "Nothing at Stake" problem and prevents short-range attacks on the system.

In the DPoS system, only the elected delegates can earn PoS revenue from running a full node. They vote for the consensus instead of the delegates themselves. The voting power of the delegates depends on their 'stake' in the network, measured by how many coins they own. The idea is that the nodes with high

---

[84] "Understanding the basics of a PoS security model", Terdermint, https://blog.cosmos.network/understanding-the-basics-of-a-proof-of-stake-security-model-de3b3e160710

[85] "Delegated Proof of Stake consensus", https://bitshares.org/technology/delegated-proof-of-stake-consensus/

stake have the best interest to maintain the functionality of the system. The voting is in real-time to achieve consensus. The elected delegates can create blocks and prevent non-trusted parties from participating. Each time they generate a block, they receive rewards as payment. The delegates can also vote out elected delegates if they do not do their job. In this way, the delegates who try to take advantage of the "Nothing at Stake" will be voted out soon enough. Like the miners in Bitcoin, they are unable to change transaction details. However, they can prevent specific transactions from being included in the next block. Such a consensus mechanism is extremely fast. To produce a block takes mere seconds.

The downside is that DPoS makes the network more centralized. The nodes with higher 'stake' can assert more control over the network. Currently, several digital currencies using PoS consensus include Peercoin, Nxt, Novacoin, BlackCoin, and BitShares.[86]

Section 4.02   PoW and PoS hybrid

A compromise solution is to develop a scheme with combined advantages of PoW and PoS schemes and to avoid their pitfalls. There are many ways to combine PoW and PoS. In one of the proposed PoW/PoS hybrid scheme, PoW is used for the distribution of new coins, while PoS is used to secure transactions. This type of blockchain is called hybrid blockchain.

How to achieve the balance between the benefits and pitfalls of the two consensus mechanisms is the tradeoff between resources and security. Hybrid PoW/PoS solution is safe against long-range attacks provided there is enough hashing power in the network. This is because mining (creation of new coins) is for long range, and minting (validation of transactions) is for short range. The inclusion of PoW blocks into a blockchain can protect against attacks. However, when a blockchain is aged, there will be more minting and less mining. The weight of PoS will be heavier.

---

[86] "Top 7 profitable PoS cryptocurrencies", Sudhir Khatwani, https://bitshares.org/technology/delegated-proof-of-stake-consensus/

Therefore, the best combination of PoW/PoS at one point may not be the best later on. Currently, there are few cryptocurrencies using the hybrid PoW/PoS. Espers (ESP) and Swisscoin (SIC)[87] are two of the examples. [88]

Another hybrid method is to alternate the use of PoW and PoS blocks. When forming a PoW block, the staking is deactivated, and when forming a PoS block, the staking is activated. The PoS blocks work as "checkpoints". This is a check and balance system. The high stakeholders check and balance the PoW miners with high hashing power. To mine PoS blocks, the stakeholders need to sign using their private keys. If there are several blocks at the same height, they can only sign one of them. In order to finalize a block, two-thirds of the validators in the active validator pool need to commit for that checkpoint. Once a block finalized, even majority of the validators working together cannot roll back the block without taking a substantial economic loss.

Ethereum developed Casper to change validation scheme from PoW to hybrid PoW/PoS. In the hybrid Ethereum system, Ethereum alternates between the PoW and PoS. In its first version, only every 100th block will use PoS consensus as a checkpoint. The checkpoint utilizes the concept of block validators. [89] Anyone depositing Ether can become a validator, who will be able to participate in the PoS consensus process. The size of a validator in the active validator pool refers to the amount of Ether that they deposited. Two-thirds of validators weighed by deposited amounts select one of the newly formed chains as the correct chain, rather than the longest chain. The selected chain is the chain containing the justified checkpoint of the greatest height. To resist majority attacks, Casper punishes deviations from the protocol by withholding rewards and locked funds from misbehaving validators. The valuator's entire deposit can be lost. Such a

---

87 "What is the hybrid blockchain? PoW and PoS explained", Chiboyanyki, https://swisscoinhub.com/what-is-hybrid-blockchain/
88 https://espers.io/
89 "What is Ethereum Casper Protocol", Ameer Rosic, https://blockgeeks.com/guides/ethereum-casper/

Consensus mechanisms

penalty is greater than the mining reward, thus PoS provides stronger security incentive than PoW.

Once the checkpoint is in place, the blocks before the checkpoints are all "frozen". PoW does not have to validate the blocks before the last checkpoint. This greatly reduces the resources required for the PoW. PoW becomes less resource intensive and can produce blocks faster. This also allows the network to scale more efficiently by partitioning a large database into smaller parts. The PoS checkpoints also make the 51% attack more difficult. At the same time, since the PoW is less resource intensive, the block reward is reduced from 5 ETH to 3 ETH. This has an added benefit of reduced inflation, which can make ETH more valuable.

As long as Ethereum uses PoW, it still has a scaling problem. Ethereum can process 10~30 transactions per second, although faster than the 4 transactions per second of Bitcoin, still pales in comparison to VISA credit card. It is still far from sufficient for large-scale applications. Ethereum is developing a sharding technique to improve the transaction speed. [90] Sharding splits the network into small partitions known as shards, with each piece containing its own independent state and transaction history. The performance of transaction speed of sharding still remains to be seen.

Section 4.03   dBFT – an alternative to PoW and PoS

There are different ways to power the blockchain besides PoW and PoS, which have their advantages and drawbacks. PoW is resources intensive, in terms of both electricity and computing power. PoS requires far less electricity and less powerful computer hardware. However, the wallet staking coins needs to be online at all times, which could be vulnerable to hackers. Furthermore,

---

[90] "Ethereum will soon test sharding tech to fix scaling issues, Vitalik Buterin hints", https://thenextweb.com/hardfork/2018/02/23/ethereum-vitalik-buterin-sharding

many people think that it is not a fair system because whoever owns more gets more reward.

At the same time, in Section 4.05, we have discussed the fork problem of PoW and PoS. If the blockchain technology becomes widely adopted in the financial sector with large-scale applications, involving billions of dollars, such as a national digital currency, stock trading, the fork problem will be amplified and most likely to disrupt the financial system and causing disasters. The financial system will not be able to tolerate this ever-lurking possibility of the blockchain splitting into two alternative versions. Furthermore, even the fastest PoS blockchain out there can accommodate a few hundred transactions per second, very different from today credit cards' transaction rate and the need for an alternative becomes clear as day. That is why some developers are searching for better solutions.

Delegated Byzantine Fault Tolerance (dBFT) based algorithm is one of these solutions. Consensus protocols for tolerating Byzantine faults provide attractive alternatives. While consensus is easy in the absence of faults, it becomes complicated when faults are present.[91]

The name Byzantine comes from a paper published by Leslie Lamport, et. al. in 1982, entitled "The Byzantine Generals' Problem" in game theory in computer science.[92,93,94] A real-life example of the Byzantine Generals Problem is the nine Supreme Court Justices during Obama administration, split 4–5 between being conservative and liberal. To achieve a consensus in the Supreme Court can be a Byzantine Generals' Problem.

---

[91] "The Swirlds hashgraph consensus algorithm: Fair, fast, Byzantine fault tolerance", L. Baird, Swirlds Tech Report, http://www.swirlds.com/ developer-resources/whitepapers/, 2016

[92] "The Byzantine Generals' Problem": Leslie Lamport et. al., https://www.microsoft.com/en-us/research/publication/byzantine-generals-problem

[93] https://medium.com/loom-network/understanding-blockchain-fundamentals-part-1-byzantine-fault-tolerance-245f46fe8419

[94] "The next 700 BFT protocols. ACM Transactions on Computer Systems", P. Aublin, et. al., 2015.

Consensus mechanisms

Leslie's paper discussed the reliability of a computer system to handle the fault caused by the Byzantine Generals' Problem that gives conflicting information. A fault is a failure or error. Failure tolerance means the system is tolerant of failure or error. In a distributed system, the faults can happen in the faulty network, crashed servers, delay in transmission, and malicious attacks. When a fault happens, the outputs can give conflicting information.

There are many ways to tolerate faults. Early isolation of the fault in a containment zone or a built-in redundancy can prevent the fault from propagating and eventually breaking down the system. The pre-requisition of doing this is an early detection and possibly early correction of the fault.

Several algorithms have been developed to make the system Byzantine fault tolerant. By using delegated Byzantine Fault Tolerance to achieve consensus, the relationship between different blockchain nodes is rearranged so that the entire network becomes almost invulnerable to the Byzantine Generals' Problem, while still being able to achieve consensus if malicious node would attempt to cause harm.

To do so, the nodes are divided into two categories: professional nodes and ordinary nodes. The professional nodes run a node as a way to gain extra income (equivalent to miners in Bitcoin blockchain). The user nodes are ordinary nodes. A delegated voting process by ordinary nodes appoints the professional nodes. This is why the name "delegated" is used.

The professional nodes have the responsibility of block verification. First, a professional node builds a block and broadcasts its version of the blockchain to the network. If two third of the other nodes agree with the information, the consensus is achieved and a block is verified. Otherwise, a different professional node is appointed to broadcast its blockchain version. This process repeats itself until a consensus can be established. This process effectively blocks the propagation of Byzantine fault. For the safety and security critical systems, it is important to prove the effectiveness of blocking the fault propagation to an acceptable level.

In view of all the potential issues of PoW and PoS, delegated Byzantine Fault Tolerance presents an attractive alternative for the blockchain application. It provides swift transaction verification times, de-incentivizes most attack vectors and upholds a single blockchain version with no risk of forks or alternative blockchain records emerging - regardless of how much computing power, or coins an attacker possesses. Ethereum's upgrade – Casper has incorporated Byzantine fault tolerant (BFT) into PoS. Byzantine consensus is extremely fast, that can process about 80,000 transactions per second with very low latency overhead.

In reality, PoW is just a variation of BFT design, with emphasis on peer-to-peer networking and cryptographic authentication.

Section 4.04 Paxos and Raft

Paxos is a consensus protocol introduced in 1989. It provides a simple algorithm to deal with the challenges of asynchrony, process crash/recovery, and message loss in a distributed system. Raft evolved from a Paxos protocol and systems providing a fault-tolerant solution to the distributed consensus problem. Both Paxos and Raft are asymmetric based consensus protocols. That is, at any given time, one server acts as the leader and is in charge and others accept its decisions. In such a system, clients communicate with the leader.[95]

In the early 2000s, Google was running into the fault induced service downtimes in its cloud-computing platform – Google File System (GFS). GFS, a Cloud-computing platform, deals mainly with big data storage, processing, and serving. Google was searching a fault tolerant coordination solution for its GFS. It adopted Paxos protocol and called it the Google Chubby. The Google Chubby project boosted interest in the industry about using Paxos protocols for fault-tolerant coordination. Since then, Paxos protocols quickly grew in number to become the protocol of choice of the consensus and coordination in the cloud-computing platform. Other industry leaders, such as Amazon and Microsoft

---

[95] "Yet another visit to Paxos", C. Cachin, Research Report RZ 3754, IBM Research, Nov. 2009.

Consensus mechanisms

also use Paxos algorithm for their cloud computing services: Amazon web service, Microsoft Azure services (Windows Fabric).

Paxos has revolutionized the distributed systems with a workable fault-tolerant consensus. However, experience in dealing with Paxos have shown that it is both hard to understand and difficult to implement. The reason is that the protocol is detached from real-world issues and use cases in the implementation. The proofs, in theory, may not hold in practice. This requires extensive proofs and verification during the Paxos implementations.

Over the time, many versions of improved Paxos appeared. The Raft is one of the most notable improvements.[96] It enhances understandability of the Paxos protocol while maintaining its correctness and performance. Consensus, in a Raft system, has a narrow definition of the assurance of the validity, agreement, and termination three properties in a distributed system.

Raft uses decomposition to improve understandability. Decomposition is the process of separating functions like leader election, log replication, and safety etc. Raft also improves the ease of implementation of the Paxos protocol by embedding the module within the replicated state machines.[97] Facebook's HydraBase uses Raft protocol.

Consensus used in the replicated state machines is different from the conventional idea of consensus, which is to find the agreed single values. Such a consensus is the single decree consensus. In a replicated state machine, the goal of consensus is to find the agreement about operations to a single replicated log. In other words, such a consensus protocol integrates several instances of single decree consensus within the replicated state machine to agree on a series of values forming the log. It is the multi-Paxos consensus.

---

[96] "Consensus in the Cloud: Paxos Systems Demystified". Ailijiang et. al.
https://www.cse.buffalo.edu//~demirbas/publications/cloudConsensus.pdf

[97] "Raft explained", Titus von Koller, http://container-solutions.com/raft-explained-part-1-the-consenus-problem/

A state machine is a program or application that takes inputs to produce outputs. When a state machine is replicated many times, it becomes a replicated state machine. A replicated state machine consists of more than two state machines, the replicated log, and the consensus module. When a client sends a command for execution in several state machines, the log of each state machine records the command. As long as the command in the log is the same and the program to execute the command is identical in each state machine, one can guarantee that the outputs will be the same from each state machine. It is the job of the consensus model to manage the logs to ensure that they are properly replicated, and determine whether it is safe to pass the command to the state machine for execution. In a minimum replicated state machine, there are three state machines. If one of them is faulty, the other two produce an identical output, which can be accepted by the consensus.

The clients interact with the replicated state machine through the consensus module. Once a command has been committed to the log, eventually, the command will apply in the same order specified in the leader's log on every live state machine.

The clients make requests to and get responses only from the leader server. This means that the replicated state machine service can only be available when a leader has been successfully elected and is alive.

Servers may crash, and messages sent to the network may be lost or delayed, so that they will arrive in different orders as they were sent. However, as long as a majority of the servers is up, the system is functioning.

Once a server becomes the leader, it becomes available to serve the clients requests. These requests are commands that are to be committed to the replicated state machine. For every received command, the leader assigns a term and index to the command, which gives a unique identifier within the server's logs and appends the command to its own log.

A fundamental difference between Raft and Paxos is that Raft implements a strong leadership. Raft integrates leader election as an essential part of the consensus protocol. Once a

Consensus mechanisms

leader has been elected, the leader will drive all decision-making within the protocol. Only one leader can exist at a single time and log entries can only flow from the leader to the followers.

Section 4.05   Proof of Concept

Microsoft announced yet another blockchain validation scheme called Proof of Concept (PoC). It is a framework for blockchain deployments in Microsoft's Azure cloud. Microsoft takes the burden of building blockchain, including the hashing and signing, so that its customers can focus on developing desirable features through smart contracts without worrying about the inner structure of the blockchain itself. By doing so, Microsoft makes the task of building web applications extremely simple. It is possible to create the blockchain application without writing any code.

The PoC framework provides users with the ability to publish underlying code and Azure services using Azure Resource Manager (ARM) templates. Users can quickly prepare Application Programming Interface (API), web application, integration, and SQL database.[98] The SQL database can be configured to collect on-chain data. The blockchain data is replicated into an off-chain store so that users can leverage their existing skill sets to enable additional capabilities. Using Microsoft's Azure Event Hubs, uses can also build additional Azure services like Azure Stream Analytics and Azure Data Lakes.

---

[98] https://azure.microsoft.com/en-us/services/sql-database/

Chapter 5   Altcoins

Since the creation of Bitcoin in 2009, Bitcoin's capitalization reached a whopping US$190 billion by the end of 2017. Nothing in the financial history has seen such explosive growth of an asset in such a short time.

Bitcoin's success has inspired many competing cryptocurrencies - Altcoins. Altcoin means alternative cryptocurrency. Today, there are over 1,000 cryptocurrencies. Altcoins try to distinguish themselves with some competitive advantages. Most of them are not very popular and operate with small market capitalizations, volumes, and limited circulation. The top four cryptocurrencies in terms of the market capitalization are Bitcoin (~US$190 billion), Ethereum (~US$45 billion), Bitcoin Cash (~US$25 billion), Ripple (~US$10 billion) and Dash (~US$6 billion) in December 2017.[99]

Different Altcoins differentiate themselves by using different hash algorithms, timestamping, block sizes, etc, or sometimes, different concept and philosophy entirely. For example, Litecoin is the first cryptocurrency to use Scrypt as a hashing algorithm. Ripple is designed as the first peer-to-peer debt transfer currency. Zcash uses Zero-Knowledge-Proof for authentication. Antshares uses delegated Byzantine Fault Tolerance in place of PoW and PoS. Other Altcoins' design tried to avoid or address Bitcoin's shortcomings.

Some of these Altcoins have their merits and may have bright future, while most of them may fade over time. We will discuss some of the most important ones, not so much from the market capitalization, but from the technology point of view.

---

[99] Cryptocurrency market capitalization, https://coinmarketcap.com/

Altcoins

Some cryptocurrencies like Ethereum and Ripple are not intended to be cryptocurrencies alone but are tokens for the enterprise solutions. They are brand new software platforms intended for the development of new applications. Their creation opens up a completely new branch of blockchain applications. In fact, their long-term impact may be much more than the cryptocurrency itself.

Cryptocurrencies are tokens in their blockchains. A token in a blockchain application means something of value to be held electronically and transferred at will without a third party involvement. Blockchain application's creator/ owner or the consensus of the participants define the rules for the transfer of tokens. For example, Bitcoin blockchain has rules to specify how many Bitcoins can be created, how many Bitcoins are rewarded to the miners executing validation of the blocks, how the Bitcoin transaction is done, etc.

In a decentralized blockchain application, such as Bitcoin blockchain, the consensus of the majority of the participants decides any change in the rules and the miner's control and executes the rules. In a centralized blockchain application, the majority of the participants and the owner of the application are usually the same ones. The participants/ owners define, control and execute the rules (such as the banks running the nodes).

Section 5.01 Litecoins – a lighter version of Bitcoin

Charlie Lee, an MIT graduate, launched Litecoin in 2011.[100] It was one of the early cryptocurrencies following Bitcoin. Its market capitalization is around US$9 billion. In every aspect, Litecoin is similar to Bitcoin - an open P2P digital currency not controlled by a central authority. The major difference is that Litecoin uses "Scrypt" as a Proof of Work, instead of the SHA256 hash function as in the Bitcoin network. A growing number of merchants are accepting Litecoin.

Scrypt uses little resources so that even consumer-grade computers are powerful enough to mine Litcoins. There are other

---

[100] https://litecoin.com/

differences: it has a faster block generation rate and a faster transaction confirmation. Litecoin platform generates a block every 2.5 minutes, instead of the 10 minutes required by Bitcoin. Therefore, the confirmation time is shorter for Litecoin. This also implies that miners receive Litecoins as reward faster than that of Bitcoin. The downside is that the Litecoin blockchain grows faster and the miners need to have more storage space than Bitcoin miners do.

The Litecoin mining equipment is less expensive than that of the Bitcoin. Anyone with a computer connected to the internet can mine Litecoins. This makes the Litecoins more decentralized than Bitcoins because the entry barrier is lower. For the same reason, Litecoin network is more prone to attack than the Bitcoin network. It would also be easier for the attacker to double-spend Litecoins compared to Bitcoins.

Faster block generation means there are less temporary forks. When a block is attached to the blockchain, all the temporary forks dissipate. On the other hand, the faster transaction time allows some miners to take advantage by generating blocks with fewer transactions to be ahead of everybody else to win the mining rewards. Like Bitcoin, Litecoin's block mining reward halves every 840,000 blocks. It started with 50 coins per block. In 2018, the coin reward is 25 coins per block. It will decrease to 12.5 coins in 2019.

Furthermore, the faster block rate also has implications for the miners' incentive in the difficulty retargeting. It can lead to the instability in difficulty adjustment. The difficulty and hash powers of the current network are two levers to control the overhead of blocks. It can mean more monetary inflation and slower transaction time.

The total number of Litecoins ever to exist is 84 million, or four times of the Bitcoin. Since Litecoin generates blocks 4 times faster than Bitcoin blocks, but the maximum number of Litecoin is also four times of Bitcoin, the monetary inflation of Litecoin follows exactly the same trajectory as that of Bitcoin, or the halving of mining rewards every four years.

Section 5.02   Zcash – a token with privacy

Altcoins

Zcash (ZEC) is a decentralized and open-source Altcoin launched in late 2016. Zcash offers privacy, extra security and selective transparency of transactions.[101,102] It claims to have the best privacy among cryptocurrencies by using Zero-Knowledge-Proof (ZKP) construction called a zk-SNARK.

Zcash derives its name from ZKP authentication. Please refer to Section 2.09 for the discussion on ZKP. The privacy option allows businesses, consumers to control who get to see the details of their transactions. ZKP construction has the advantage of ensuring the validity of transactions and securing ledger of balances without giving out any other information. Bitcoin lacks this feature.

Most of the cryptocurrencies do not hide the amount of the transaction and the parties involved in the transaction. The parties involved are represented by their addresses. However, in the exchange, your addresses are linked to your bank accounts. Therefore, privacy is not guaranteed.

Some cryptocurrencies, such as Bitshares, can hide the transaction amount, but not the receiver and recipient. Others, such as Monero, create a large number of dummy addresses as well as inputs and outputs for each transaction to hide the identity. However, it is not 100% privacy proof.

Zcash uses cryptographic zero-knowledge proofs to protect both the amount and recipient of shielded transactions. Zcash also allows for transparent transactions, which do not provide any privacy protections at all.

Section 5.03   Ripple – a digital equivalent of SWIFT

Ripple is not a cryptocurrency in the traditional sense. Ripple Labs, the founder Ripple, created 100 billion XRP at

---

[101] https://z.cash/

[102] "Zerocash: decentralized anonymous payments from Bitcoin", Eli Ben Sasson et. al., Zerocash white paper, http://zerocash-project.org/media/pdf/zerocash-extended-20140518.pdf

Ripple's inception in 2012.[103] It releases XRP over time according to its release strategy. As of early 2018, there are 39 billion XRPs in circulation. It operates based on a Ripple Transaction Protocol (RTXP) as a real-time global settlement network that offers instant, secure and low-cost international payment system. [104]

The development of Ripple payment protocol actually predates the development of Bitcoin. Ryan Fugger[105] proposed the system in 2004. He intended to create a decentralized monetary system that could empower individuals to create their own money.

All money in Ripple is either debt or credit line. Transactions simply consist of shifting balances from the payer account to the receiver account. This is very much like the bank debit card and credit card system. Only the people who had a "credit line" in the Ripple system could join the network, and so one could trust each other to pay each other back as needed. [106]

Ripper works similar to the SWIFT,[107, 108] an acronym for the Society for Worldwide Interbank Financial Telecommunications, which is a system that handles the transfer of money internationally. [109] Actually, SWIFT is a messaging network that financial institutions use to transmit instructions of money transfer through a standardized system of codes.

When one wants to wire money to another person in a different country, he goes to the bank to deposit the money and

---

[103] https://ripple.com/
[104] "Ripple: Overview and outlook." Proc. Trust and Trustworthy Computing", F. Armknecht, et. al., volume 9229 of Lecture Notes in Computer Science, Springer, 2015.
[105] http://ryanfugger.com/
[106] Vitalik Buterin, "Introducing Ripple": https://Bitcoinmagazine.com/articles/introducing-ripple/
[107] https://www.swift.com/about-us
[108] "The Society for Worldwide Interbank Financial Telecommunication", S. V. Scott, M. Zachariadis, www.oapen.org/download?type=document&docid=623230
[109] "Ripple vs. SWIFT: payment revolution, David Blair, Treasury Today, http://treasurytoday.com/2017/07/ripple-vs-swift-payment-r-evolution-ttpv

Altcoins

gives the instruction, which includes the transfer amount, the recipient's name, address and bank account information. The bank then sends a message to the destination bank, to advise such a wire. The recipient can withdraw the money from the destination bank. In reality, the bank does not transfer the money. The bank only sends a message describing the transfer. The two banks settle their transactions later. Ripple works the same way. The party who interface with the customer is the gateway. Ripple records the transactions in its distributed ledger.

Apparently, this type of transaction requires trust. In the Ripple system, there are agents of trust to provide the trust. If the transaction parties do not have mutually known trust agent, Ripple will find a trust agent acceptable by the parties of the transactions. The Ripple algorithm tries to find the shortest trust path between the gateways.

By using such a chain of trust, money can change hands even between people who do not know each other at all. For the first few years, only a very small community used the Ripple system. This is because, without blockchain type of technology, the system allowed only the trusted people to join the club. In 2012 when the system adapted blockchain technology with the introduction of a cryptocurrency called Ripple (XRP), the trust problem was resolved, and the system expands worldwide.

Ripple introduces the concept of "gateway", which is a link between the XRP ledger to the outside world. Ripple enlisted financial institutions to be the gateways. The gateway also serves as the link in the trust chain to make a transaction. Since there are many gateways to choose from, the system still preserves a measure of decentralization comparable to mining pools. Only when the assets move in and out of the Ripple network, the gateways convert XRP to fiat currency and vice versa. Within the Ripple network, all transactions take place in XRP. In this way, Ripple operates like a national domain, where XRP is its currency. The gateways are like customhouses. Only when the external transactions are involved, XRP is converted into other currencies.

The Ripple network keeps a public ledger of accounts, balances, and IOUs. The network constantly updates the ledger and distribute to all the servers in the Ripple network around the world. There are two kinds of assets in the Ripple network: XRP and issuance. Issuance is the digital balance that represents asset held by an issuer. A gateway sends issuance to the address of the customer, the issuer, who deposits an asset or money to the XRP Ledger from outside of Ripple network. Thus, issuance is like a receipt of a deposit of an asset. XRP is a coin to trade issuance in the Ripple network. When the issuer decides to send the asset to someone outside of the Ripple network or withdraw it, the process goes through a gateway. Figure 5-1 shows the Ripple network, its gateways, and assets in the network.

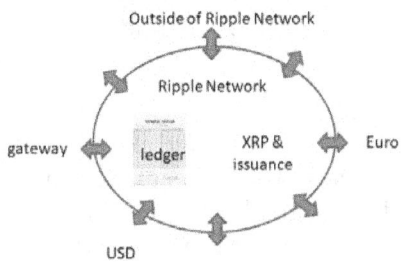

Figure 5-1 A Ripple network

Transactions in XRP settle immediately, so does the Ripple network update the ownership of the actual asset changes. The transaction verification is by consensus among members of the network.[110] Transactions in Ripple incur smaller transaction fees paid in XRP. The servers agree on changes by consensus. Anyone can be a server by running free software on his or her computer. The Ripple rules govern the transactions. There are total 100 billion XRPs.

Even though Ripple is a debt based system; XRP is not debt- based cryptocurrency. One can send XRP from one Ripple address to another over the Ripple network just like Bitcoins in the Bitcoin network.

Like Bitcoin, Ripple operates as a decentralization system. There is no central authority to validate the transactions. Ripple recruited a list of organizations to validate the transactions. Among them are Microsoft, MIT, a Swedish ISP Bahnoff, CGI,

---

[110] "The Ripple protocol consensus algorithm", David Schwartz, et. al. https://ripple.com/files/ripple_consensus_whitepaper.pdf

Altcoins

WorldLink, ST TOKYO Corporation, and many others. Currently, there are roughly 166 validators in Ripple's validator registry, which lists their public keys. [111] The non-validator nodes are called the monitoring nodes.

Its mechanism for keeping track of balances is similar to Bitcoin. Transactions are done using addresses, public and private keys, and modifications to the database through digital signatures. Ripple private and public keys use the same elliptic curve cryptography as Bitcoin. Elliptic curve cryptography (ECC) is a cryptography based on the algebraic structure of the elliptic curve over finite fields.[112] One can use the same private and public keys to sign transactions and messages in the Bitcoin and Ripple networks.

Ripple does not use PoW or PoS for its consensus. Individual nodes decide which version of a new ledger to accept by polling the nodes around them to see what the majority opinion is, allowing the network to settle on a single choice. The process is much faster than Bitcoin block confirmation.

This consensus approach has a risk of causing chain fragmentation, where two parts of the network settle on irreconcilable transaction histories. It is vulnerable to the attack, which employs various forms of proxy and IP address spoofing techniques to pretend to be a million separate nodes, and so overwhelm the opinion of the rest of the network through sheer numbers.

To avoid this risk, Ripple introduced a Unique Node List (UNL). [113] Instead of taking the consensus from the wider, unknown network, the UNL of a node provides the consensus. The Ripple Protocol Consensus Algorithm (RPCA) makes UNL flexible. Each node can choose its own trusted nodes in its UNL. Different nodes have different UNLs. A typical UNL contains

---

[111] https://xrpcharts.ripple.com/#/validators

[112] "Elliptic Curve Cryptography: a gentle introduction"", Andrea Corbellini, http://andrea.corbellini.name/2015/05/17/elliptic-curve-cryptography-a-gentle-introduction/

[113] "Unique Node List", https://wiki.ripple.com/Unique_Node_List

100+ nodes (i.e. their public keys). Ripple provides a recommended list of nodes which can serve in a UNL.

There is a requirement that the pairs of validating nodes need to have a minimum number of nodes in common or overlapping. Otherwise, the ledger would fork. Ripple states that the overlap should be at least 1/5 of the size of the larger list. Since Ripple is a trusted public network, most nodes will connect to probably 10-100 large organizations running validators, and these organizations will connect to each other. This makes a split reasonably unlikely.

By using UNL, essentially Ripple duplicates the original confined trust environment million times over the Ripple network. To ensure the trustworthiness of the UNLs, Ripple platform runs auditing software to analysis the UNLs and report on potential degenerate lists so that they can be corrected.

A node chooses to accept the ledger which is validated by the consensus of its UNL, under the assumption that the nodes are not likely to work together to push a fraudulent ledger by consensus. Once the nodes reach the consensus, all nodes update their ledgers to the consensus ledger.

The UNL system also ensures that the network is tightly linked. Every node connects to every other node in millions of ways, and so all nodes are only at most a few hops away from each other. Thus, any fragmentation would rapidly resolve itself.

Ripple builds ledger in a ledger chain differently from Bitcoin's block in the blockchain. Instead of keeping all the transaction history, Ripple ledger consists of a transaction tree and a state tree. The transaction tree shows the transactions that have taken place since the last ledger. The state tree contains all of the information of the account balances and credit limits in the Ripple system.

The Ripple ledger does not carry the transaction history beyond the last ledger. This greatly simplifies the information to be stored in the ledger chain. The node also needs to deposit a minimum balance of 200 XRP for creating an address and an additional 50 XRP for creating a credit line. This is a strong disincentive against bloating the ledger state with many addresses. As a result, the majority of Ripple clients are fully participating

Altcoins

nodes, as the cost of full participation is small enough to be negligible for most computers. This ensures a greater degree of decentralization.

The lack of mining in Ripple is both strength and weakness in the system. Without mining, nodes cannot receive a reward by doing the work. Instead, Ripple platform started with all 100 billion XRP that will ever exist. The developer of Ripple, OpenCoin, pays transaction fees out of the 55 billion XRP reserves it is holding.

Without mining, XRP is deflationary, because the existing XRP unit can only decrease but not increase. The number of XRP starts at 100 billion and then gradually decreases some XRPs are lost. This means XRP coin price can increase over time.

One can trade anything, e.g. USD, CAD, gold or even airline miles in the Ripple network. For example, a seller of airline miles can deposit miles credit at a gateway. A buyer deposits USD at another gateway. The trade is carried out in the Ripple network.

Transactions on a distributed network with public ledgers is faster, cheaper, lower risk, and much, much better in almost every way possible than centralized pre-Internet correspondent banking messaging networks such as SWIFT.

Ripple also has the potential for much greater integration with the existing banking system, as its currency exchange is a service that even existing financial businesses will quickly be able to benefit from. Many large banks are testing XRP in blockchain trial.[114] For those who are afraid of the prospect of gateways defaulting or disappearing, or who wish the greater privacy that Bitcoin's trust-free anonymity offers, Bitcoin continues to be the best bet.

Regardless, the fact that Bitcoin now has a strong and compelling alternative makes it clearer than ever that the idea of cryptocurrency as a whole is here to stay. Ripple's technology can connect the world's ledger and networks without replacing the

---

[114] "Global Banks Test Ripple's Digital Currency in New Blockchain Trial", Michael del Castill, https://www.coindesk.com/global-banks-test-ripples-digital-currency-new-blockchain-trial/

existing systems. It gives financial institutions full control and privacy over their transactions through RTXP.

Section 5.04 Ethereum: the smart contract blockchain

Ethereum deserves more attention because it is the second most popular cryptocurrency behind Bitcoin. [115] It also distinguishes itself for being the first Blockchain platform to run smart contracts. Smart contract is the coding on the blockchain that allows transactions to take place. It can help deal with the uncertainty and complexity of the real world, while enhancing reliability.

Today, Ethereum is the largest and most well established, open-ended decentralized software platform. It has the second largest market capitalization of crypto currencies after Bitcoin at $45 billion in early December 2017.

Vitalik Buterin and the Ethereum Foundation launched Ethereum platform on July 30, 2015. Ethereum is a distributed blockchain platform that enables to build and run Distributed Applications (DApps)[116]. DApps execute smart contracts in the Ethereum blockchain. It is secure, free from downtime, fraud, control or interference from the third party. The smart contracts automate many of the procedures that require human intervention today. Currently, there are over 6,000 computers in the Ethereum blockchain network. Every computer in the Ethereum blockchain network executes the DApp codes.

The cryptographic token Ether pays the service of execution of DApps on Ethereum. In 2014, Ethereum launched a pre-sale of Ether with great success. The blockchain also allows people to the transaction in Ether.

Besides Bitcoin and Ethereum's commonality in the concept of distributed ledgers and cryptography, they are different in many fundamental ways. The major difference is their

---

[115] https://www.ethereum.org/
[116] "What are Dapps? The new decentralized future", https://blockgeeks.com/guides/dapps/

purposes: Ethereum is not just a platform but also a blockchain based programming language, which allows developers to build and publish distributed applications (DApps). DApps to Ethereum blockchain is like apps to the Android operating system on the cell phone.

Ethereum, as a platform, facilitates peer-to-peer contracts and applications via its own token. Anyone can go to MyEtherwallet.com and click the tab "Contract".[117] There you can deploy a contract by entering your own code. This is not possible with Bitcoin.

The applications of Ethereum are not limited to the peer-to-peer network; they run in the cloud-computing environment as well. This greatly enhances its commercial applications. Ether can codify, decentralize, secure and trade just about anything. This capability created a software foundry business. Similar to the semiconductor foundry which uses IP's to build application-specific circuits for customers; software foundry uses Ether to build application specific software for its customers. Companies interested in building Ethereum applications but without know-how or resources can outsource the project to software foundries.

The startup ConsenSys,[118] based in New York, offers Ethereum Blockchain as a Service (EBaaS) in early 2015 as a software foundry to develop decentralized services and applications that operate on the Ethereum blockchain both in the cloud and in the peer-to-peer network.

Even software giant Microsoft formed a partnership with ConsenSys to offer customers of the Microsoft cloud-based enterprise computing service platform, called Azure, to develop cloud-based blockchain applications using EBaaS.[119]

The combination of cryptocurrency with distributed applications differentiates Ethereum from Bitcoin drastically with

---

[117] https://www.myetherwallet.com/#contracts
[118] https://consensys.net/
[119] "Ethereum Blockchain as a Service now on Azure", Marley Gray, https://azure.microsoft.com/en-us/blog/ethereum-blockchain-as-a-service-now-on-azure/

huge implication. Ethereum brings the blockchain technology to the cloud computing. In doing so, Ethereum has created a massive decentralized computing platform that merges the cloud with a peer-to-peer network. Many people call it a "world computer."

Ethereum is both a cryptocurrency and a software platform. It is also a utility token. It trades like any cryptocurrency, and yet it powers the creation and execution of smart contracts. In real life, you can find many analogs. For example, petroleum can be traded as commodity, but it can also power automobiles. In fact, Ethereum works as any kind of commodity. You can buy and sell Ether the same way as you can buy and sell coal. Coal can power steam engine, as Ether can power smart contracts.

Ethereum has created a rapidly growing ecosystem of software startups vying to build decentralized applications, known as "DApps". These DApps create markets, store registries of debts or promises, move funds in accordance with instructions given long in the past like a will or a futures contract, and many other things that have not been invented yet, all without a middleman or counterparty risk.

The programming language used by Ethereum is the Ethereum Script, also known as EtherScript. It writes agreements without ambiguity and is more precise than the agreements written in words. A smart contract written in the script can precisely define the terms and conditions governing the transfer of value and assets and automatically enforce the execution of the contract with the same script. The cryptocurrency Ether serves as the transfer media.

There are many other less noted differences between Bitcoin and Ethereum. For example, the block time in Ethereum transaction in seconds rather than ten minutes as in Bitcoin. Their hash algorithms are also different.

Microsoft is not the only large company interested in working with Ethereum. Other large companies across different industries such as JP Morgan Chase, Cisco, Bank of New York Mellon, Red Hat, CME Group and Banco Santander are working with Ethereum in one way or another, so as the Chinese internet

Altcoins

giants: Alibaba Cloud, Tencent, and Wancloud. [120] In fact, the Enterprise Ethereum Alliance (EEA) boasts a membership over 150 companies and organizations and is becoming the largest open-source blockchain initiative in the world.

Ethereum and Bitcoins are at the two opposite ends of the centralization and decentralization axis. Ethereum Foundation, the creator of Ethereum platform, guides, and controls the development of Ethereum applications. This is opposite from the creator of Bitcoin platform, the Bitcoin Core, which cannot force any change in the Bitcoin platform to the Bitcoin community.

In the Bitcoin platform, the majority of the Bitcoin community needs to reach consensus to implement any change. As a result, Bitcoin community can hardly reach an agreement on many technical issues. This is probably also the reason that Ethereum, having a shorter history, has already implemented more hardforks than the Bitcoin.

Less decentralization, however, does not diminish Ethereum's importance in any way. Because the applications built by using Ethereum platform is still decentralized. In a way, the decentralization takes a different meaning when referring to Ethereum.

In 2016, Ethereum had a hardfork. The token split into Ether Classic (ETC) and Ethereum (ETH). We will discuss the hardfork event in more detail in Section 5.05. The cap for Ethereum Classic (ETC) issuance is roughly 230 million ETC's, a supply much higher than that of Bitcoin. The block reward reduces by 20% for every 5,000,000 blocks created.

Ethereum (ETH), on the other hand, takes a different approach to control the ETH reward. Instead of reducing the mining reward, as in the case of Bitcoin, it increases the block time by design. The block time started to increase in mid-2017, and the increase is exponential. Starting from Block 3.5 million,

---

[120] "9 Brand-Name Companies That Have Joined the Enterprise Ethereum Alliance", Sean Williams, The Motley Fool, https://www.fool.com/investing/2017/07/27/9-brand-name-companies-that-have-joined-the-enterp.aspx

the next 100,000 blocks will have average block interval 25 seconds. Thereafter, the block interval increases to 35 seconds, 55 seconds, 95 seconds, etc. for every 100,000 blocks created, and so on. The asymptotic limit will occur around 2021 when the network will be almost frozen."[121]

However, the mining reward does not change. This has the same effect as capping the supply because it takes longer to mine the Ethereum. When the block time is doubled, the supply of ETH through mining per fixed period is halved.

In this way, the total supply of ETH will be capped around 100 million. However, we should note that the Ethereum Foundation could change the cap in the future. Such a change in the Bitcoin supply requires the vote from Bitcoin community, while in Ethereum, it is more centralized and be decided by the Ethereum Foundation. This is another difference between the Bitcoin and Ethereum in term of the decentralization. When Ethereum Foundation decides to make a change in the protocol, and if part of the Ethereum community does not agree, they can depart the Ethereum blockchain and cause a hardfork.

In Ethereum's short history, there are already five hardforks. There are more changes to come. According to Ethereum's roadmap,[122] the development of Ethereum has four stages: Frontier, Homestead, Metropolis, and Serenity. Each stage adds new features and improves the user-friendliness and security of the platform.

Out of these changes or upgrades, Metropolis receives more attention. Metropolis offers significant improvements to the Ethereum ecosystem. The implementation of Metropolis occurs in two phases designated as Byzantine and Constantinople. Some of the improvements are for the developers. The major improvements

---

[121] "Ethereum is Entering the Ice Age", Jose Breslauer, Steemit.com, https://steemit.com/ethereum/@joshbreslauer/ethereum-is-entering-the-ice-age

[122] "Ethereum roadmap", https://steemit.com/cryptocurrency/@ctyptouniverse/ethereum-roadmap

Altcoins

include ZK-Snarks (Zero-Knowledge Succinct Non-Interactive Argument of Knowledge), PoS/PoW hybrid, Revert, Returndata, and Account abstraction.

zk-Snarks is meant to improve the transaction privacy. This feature is already available in the Altcoin called ZCash. Ethereum developers are working together with the ZCash team to implement zero-knowledge proofs. We have discussed the PoW/PoS hybrid in Section 4.01 and Section 4.02.

The Revert and Returndata options allow the smart contracts to go back to the original state without eating up all the Gases. Gas is the cost of running Ethereum Virtual Code (EVC) or commands to execute a transaction or contract.[123] The Gas system is akin to the KWh for the electricity used to run an engine. The necessity of using Gas instead of Ether to pay the cost of running EVC is that the price of charging the service of running EVC will not inflate if Ethereum increases its value. Imagine that when the ETH price in USD increases by over 1,000% in 2017, running EVC will be highly profitable if it is priced in ETH.

A miner requires receiving an amount of Gas to perform an operation for Smart Contract according to the amount of work. The amount of Gas used depends on the smart contract. The exchange rate of Gas to Ethereum also fluctuates. The Gas system is like that the household electricity bill can change both because of the usage in KWh and the unit price of electricity per KWh in dollars.

In the Gas system, the smart contract (or the user) sets the amount of Gas rather than by the miner. Bitcoin miners prioritize transaction with the highest mining fees. The same is true of Ethereum. If the user sets the gas price limit too low, no miner will be interested in picking up such a smart contract.

The Gas price per transaction or contract is set up to deal with the Turing Complete nature of Ethereum and its EVM (Ethereum Virtual Machine Code). If there is not enough Ether in the account to perform the transaction or smart contract, then the

---

[123] "What is the gas in Ethereum", https://www.cryptocompare.com/coins/guides/what-is-the-gas-in-ethereum/

transaction will not be processed. This can prevent attacks from infinite loops, to encourage code efficiency and to make an attacker pay for the resources they use, from bandwidth through to CPU calculations through to storage.

Before the Metropolis upgrade, if the Gas limit was set too low and not acceptable to the miner, the gas will still be used. This is as if you hire a contractor to repair your garage door if you find that the quote is too high and decide not to use the service, you will still have to pay for his visit. With the upgrade, the visit does not cost anything. Best of all, the Returndata option allows the contract user know exactly why his contract fails.

Account abstraction allows users to define their wallet address, or the private key, in the form of a smart contract. By doing so, private keys to control external accounts would be less susceptible to attacks against the signature scheme. It also adds other security schemes, such as hash ladders. Abstraction also allows contracts to pay for Gas. The more complex contract pays more Gas for execution. For example, the price to pay for a SHA3 (Secure Hash Algorithm 3) operation is 20 Gases. However, this price is not fixed. It can change due to demand and supply.

There are many other improvements. For example, Light clients are important to many users, who do not want to run a full wallet. Some prefer fast blockchain synchronization. Metropolis will make light clients more secure.

There will be further new technical features in Metropolis through continuous updates. Smaller updates can be added to the protocol easily. The features, which have the potential to cause hardfork, have the priority. In 2017, these priorities are scalability, safety, and privacy.

Casper is the software update for the Metropolis. Its developers publish the smart contract on Ethereum, creating an official Casper account where anyone can deposit Ether if they want to engage in the virtual mining process.

Since Ethereum blockchain is more an application platform than a cryptocurrency, it has multiple development tools for the developers: a Python-based Ethereum client called Pyethereum, C++ based Ethereum client called Cpp- Ethereum and Go based Ethereum client called go-Ethereum or goeth. The client allows one to communicate with Ethereum network to create accounts, commit transactions and do many others. Using these tools, developers can quickly develop DApps for their particular use.

Section 5.05　　DAO hack and Ethereum fork

A German company called Slock.it (the name implies Smart Lock) used Ethereum platform to create an app called the DAO (short for Decentralized Autonomous Organization). Among many ideas to connect the IoT, Slock.it intends to build "smart locks" that let people share their properties (cars, boats, apartments) in a decentralized version of Airbnb. They call it the Universal Sharing Network.

DAO, an Ethereum app, was their first project. It runs on Ethereum platform and uses Ether as its transaction currency. At that time, DAO was the largest app on the Ethereum platform. Therefore, it gained substantial support from Ethereum Foundation. Slock.it designed the DAO to automate organizational governance and decision-making. Individuals can work collaboratively on the DAO platform. A registered corporate entity can also use DAO to automate formal governance rules contained in corporate bylaws or imposed by law.

The vision of DAO is to create organizations, in which the participants maintain control of contributed funds and the governance rules are formalized, automated and enforced using the software. Think DAO as an automated tool to run venture capital. On one hand, DAO raises funds from investors through crowdfunding event. On the other hand, DAO solicitants startup companies to present their projects, which seek investments. The crowdfunding participants then vote for the projects they want to invest. The whole process is conducted on the DAO's Ethereum based blockchain using smart contracts.

A DAO project becomes active by its deployment on the Ethereum blockchain. Once deployed, a DAO contract requires Ether to engage in transactions on Ethereum. Like Bitcoin, the DAO is open source. The code runs across a network of independent machines, and anyone can change the code if approved by a majority of the DAO's "voting power", which is in proportion to the money a voter has invested.

In the first DAO project, 50 project proposals are to be voted for funding. The money raised by a crowdfunding event organized by DAO will fund the projects selected by the vote. DAO passed the audit by Déjà vu Security on April 5, 2016, in preparation for a crowdfunding event.

On April 30, 2016, Slock.it launched crowdfunding with 28 days funding window. It was a phenomenal event. By May 27, 2016, the deadline for investing in the DAO, over ten thousand people had anonymously poured more than $168 million into this new online creation. That makes it the largest crowdfunded project ever.

The way it works is that investor buys DAO's tokens with Ethers he owns. In effect, investors invest in DAO's crowdfunding event by their Ethers. The price of the token in Ether fluctuated with supply and demand. Think Ether as money and token as stock price. DAO's token grants its holder voting and ownership rights.

When voting starts, the investment fund is frozen or locked-in. Token ownership is freely transferable on the Ethereum blockchain only after the crowdfunding event. Each account has one vote. The weight of the vote is proportional to the token held in the account.

However, DAO's voting process has an inherent logic flaw. Instead of allowing each account owner to allocate his tokens to each individual project he would like to invest, all the tokens in DAO are lumped together. The only control investor has is to cast his vote for each project. If the project is approved, investor's fund will go to the project, even if he voted for NO. Investors do not have control over how much funds to each project or whether he wants to commit to the project or not because the funds are allocated to the projects by DAO. In this sense, the token is more like stock price of the Exchange Traded Fund, which consists of a basket of stocks. Buying a token is like buying a basket of projects.

As the event unfolds, in June 2016, a security bug was discovered and disclosed to the public. One week later on June 17, DAO was hacked, most likely due to the disclosure. The hacker drained 3.6 million Ethers to a child DAO. The market value of

Altcoins

the siphoned Ether was over 70 million dollars. It caused a panic in the market and the price of Ether dropped to half.

In reality, the attacker did not break into the code, but rather he explored a weakness in the DAO rules to transfer the funds to another DAO controlled by the hacker. To the Ethereum platform, it merely executed a contract.

Fortunately, per DAO coding, there is a waiting period of 28 days before the fund can be spent, so that the fund was temporarily safe for 28 days. In order to protect the fund after the waiting period, the Ethereum Foundation froze the fund. The owner of Ethers still could see their Ethers in their accounts, but cannot withdraw.

On June 24, Ethereum Foundation released a softfork client intending to censor transactions from the hacker, however, because the hasty development, the software contained a flaw and miners decided not to implement. It seemed that a hardfork solution was necessary, as hot debates going on in the Ethereum community.

The hardfork spec was finally announced on July 15, 2016. The hardfork effectively reversed the hack from occurring and returned all funds to the original investors. However, not everyone was happy about the hardfork decision. Some felt that the hardfork violated the spirit and intent of a truly decentralized system adhering to the virtues of irreversible transactions.

By July 20, 80% of the nodes voted to approve the new client. The remaining 20% did not. The dissidents decided to continue to validate the old version of the blockchain, with the hack still present. Ethereum was therefore split into Ethereum Classic ETC and the new Ethereum ETH. The ETC community left Ethereum Foundation and is now run by a team called IOHK.

ETC started trading in the exchanges on July 23, 2016. As Ethereum classic is a replica of the original blockchain, except for a few key changes regarding the DAO transaction reversals, everyone who had tokens on Ethereum at the time of the fork now has the same amount of tokens on Ethereum classic. This is the same as a stock spin-off. The only thing that Ethereum token holders should be aware is they can be affected by "replay attacks" if they don't properly "separate" their addresses to differentiate them on each blockchain.

By August 31, Ethereum Foundation had unlocked the DAO frozen funds and started to refund some Ethers to the owners. After the hardfork, three DAO refunds took place for investors who still held DAO tokens. The last refund is the DAO to ETC contract that gives balance holders until April 15, 2017, to claim their reimbursement. After the deadline, there are still 1,717,513 Ethers remaining to be withdrawn.

The DAO hack may be a setback for Ethereum. However, the incident is not due to bugs in the Ethereum network. The Ethereum network, which supports around $1 billion worth of Ether, has never been hacked. However, by strongly supporting DAO, Ethereum's name also has been tainted by the hack. At the time of the hack, DAO had roughly 15% of all Ethers. This fact alone is enough to crash the price of Ether. Ethereum's action is akin to a bailout. Whether Ethereum took or did not take the action, it suffered the consequence.

Section 5.06   Legal issues

The use of blockchain, DAOs, and smart contracts raises significant legal questions. As the technologies become more widely used, it becomes urgent for legislators, regulators, and courts to provide a proper legal framework for the blockchain applications. However, to establish the legal framework for blockchain is overwhelming. Many circumstances that never existed before can occur. What are the jurisdictional and applicable laws to address a breach or failure occurred in a server located in a remote country? DAO is self-governing software engaging in commerce, what legal status, or liability it or its creator has. Can the law legally enforce the smart contracts or more specifically their codes? What if a software bug in the codes causes the fault? After all, the coding may suffer from errors, or the hosting platforms may fail. This is a small revelation in the large context of legal issues involving artificial intelligence

Altcoins

including the self-driving cars. A woman was dead after a fatal crash involving a self-driving car in Arizona, in March 2018.[124]

The impacts of fraud at any point in the DAO's creation or operation can be very real. Courts and regulators across the world are unlikely to allow the wholesale adoption of technology before establishing the legal framework. There are a number of possible ways to approach the issues. The easiest solution is to embed DAO into a real contractual agreement spelling out the prevailing terms for the linked DAO. A contract can spell out provisions to include or exclude certain potential faults. In this way, the contractor has the responsibility to a certain fault caused by DAO. The contracting parties can also adopt a free-to-use platform with an agreed code. The legal framework can treat DAO itself as a tool, such as a lathe machine, the creator is responsible for certain accidents but not all accidents. This way essentially separates the blockchain infrastructure from the contractual agreement and regards the DAOs or smart contracts as tools of the execution.

The adaption of common law to technological changes is not a new event. Legislatures will have to consider what legal status to grant to DAOs before the wholesale adoption of this technology.

Section 5.07 DApps – Decentralized Apps

DApp stands for Decentralized App. The most popular platform to build DApps is Ethereum. Currently, there are more than 900 DApps either in deployment or in development.[125] Its growth is quite astonishing considering its short history of fewer than 3 years.

Just to mention a few "live" ones:

- Storj: a decentralized cloud storage
- Numerai: a hedge fund
- Pick 3: a lottery draw

---

[124] http://money.cnn.com/2018/03/26/technology/arizona-suspends-uber-self-driving-cars/index.html
[125] https://www.stateofthedapps.com/

- Share & Charge: a platform for sharing electric vehicle refueling station
- Icebox: a cold storage for Ether
- VDice: on chain Ethereum betting game

Creating DApp does not require copyright or patent. Anyone can create DApp if he/she knows how to do so. Create DApps in the Ethereum blockchain ecosystem is similar to the creation of apps for the smartphone. However, there is a difference that DApps can issue tokens while apps in a cell phone cannot. Through tokens, DApp developers can directly monetize their development effort.

DApps have the following characteristics: The application must be completely open-source, it must operate autonomously, with no entity controlling the majority of its tokens, and its data and records of operation must be cryptographically stored in a public, decentralized blockchain.

The application generates tokens according to a standard algorithm or set of criteria and distributes some or all of its tokens at the beginning of its operation. The participants need tokens to use the application and earn tokens for their contribution as reward payment. There are several ways to obtain tokens:

- Crowd-sale: to buy tokens during the initial DApp crowdfunding.
- Participation in the Development: to earn tokens by doing the development.
- Mining: to receive tokens by contributing resources.

Any change in improvements and market feedback is decided by majority consensus of its users. These contributors, whether miners or developers, benefit from the exchange of the tokens and from the possible appreciation of their value. Therefore, DApps are self-sustaining because they empower their stakeholders to invest in the development of the DApp

Over time, DApps will see self-improvement and increased functionality. With the crowd wisdom, many DApps may become even more powerful than the services provided by large corporations today.

Section 5.08 Create your own coins out of Bitcoin

The Counterparty is a platform for creating peer-to-peer financial applications on the Bitcoin blockchain. It aims to bring the smart contract capability of Ethereum to the Bitcoin blockchain. It established the world's first functioning decentralized digital currency exchange, as well as the ability to create their own virtual assets, issue dividends, and create price feeds, bets, and contracts for difference.

Unlike Ethereum, another smart-contract platform, Counterparty does not have its own blockchain. Counterparty's developers built a distributed financial system on top of the Bitcoin blockchain. Since it does not have its own blockchain, it cannot create its own token by mining or minting. Rather, its token, called XCP, is created by "burning" Bitcoins. In January 2015, Counterparty announced the Proof of Burn concept to create XCP. XCP is used to pay for the execution of smart contracts in the Counterparty network.

The process works like this: You take 50 of $20 dollar bills to the US Treasury Department, and US Treasury Department burns it, and mints 10 of the $100 dollar bills for you.

The burnt Bitcoins are sent to the address provided by the Counterparty, 1CounterpartyXXXXXXXXXXXXXXXXUWLpVr, which is guaranteed that it has no matching private key because the middle "X"'s are left to chance. These "burned" Bitcoins are eliminated from the Bitcoin circulation forever.

Counterparty issued XCP to whoever submitted the proof of burn. Counterparty itself also burned 2,125.63 Bitcoins in January 2014 (worth roughly $1.8 million at the time). Up to September 2015, only 2,673 Bitcoins have been lost, for 2,648,755 XCP's. It is a small fraction (0.02%) of the total 21 million Bitcoins that will ever exist. It does not have any impact on the supply of Bitcoins. This ensured that all 2,648,755 XCP that will ever exist were created and distributed in a fair, transparent and public manner.

Like Bitcoin, the Counterparty community operates in a decentralized manner where everyone has an equal say in the project. The Counterparty is a second-level protocol that runs on top of the existing Bitcoin network and pays Bitcoin miners small fees to register Counterparty transactions in the Bitcoin blockchain. In this way, Counterparty benefits from having a trusted and secure mining network without spending any resource for its operation. At the same time, Counterparty developers eliminated any speculation that they planned to get rich quick or redistribute risk unequally.

The Counterparty is the first (and so far the only) protocol to have a working distributed exchange, built in record time despite having no outside funding of any kind. Since then, Counterparty has built a new and exciting level of functionality in the form of a distributed financial system and exchange on top of Bitcoin. That is many potential benefits for the Bitcoin community as a whole in return for those burnt coins!

The protocol specification and all Counterparty software is open source. To ensure the success of Counterparty, they have developed Counterwallet, a secure web wallet for managing all transactions on the world's first peer-to-peer digital asset exchange. Counterwallet is based upon an escrow system to keep all transactions safe and secure and to eliminate the intermediary, necessary in a typical market system. It provides a secure, safe, and open-source way to manage the money you use in Counterparty.

Counterparty allows users to issue assets, including tokens, which is a user-created token. User-created tokens are just as real as XCP or even BTC. With the asset issuance function, every user has the ability to create a new currency project inside the Bitcoin and Counterparty ecosystem. They are separate from Bitcoin the currency itself but exist entirely inside ordinary Bitcoin transactions.

These tokens work the same as Bitcoins. Any Bitcoin address can receive, store, and send these tokens. Likewise, Bitcoin cold storage can also store these tokens. However, Counterparty tokens do not count as BTC balance in the Bitcoin address. This means that sending/receiving Bitcoins has no effect on the balance of tokens and vice versa. Among other features, Counterparty adds the ability for users to create, send, trade, and

Altcoins

pay distributions on assets, in a fully decentralized and trustless manner.

While Counterparty has its own internal currency (XCP), trading and creating assets does not require anything apart from regular Bitcoin transaction fees. To transfer assets to the Counterparty, it has to contain four important parts: the source of the asset, name of the asset, quantity of the asset and asset destination.

Counterparty protocol allows the source to transfer issuance rights of the asset. Moreover, an asset can also be locked, so that there can be no further issuances of it. Beyond creating the most basic asset, it is also possible to make assets either divisible or callable. If an asset is deemed to be divisible or callable upon its initial issuance, it must always be divisible or callable with every issuance thereafter. A divisible user-created asset is, like, Bitcoin and XCP, divisible up to 8 decimal places. A callable asset is an asset which the issuer can call back (i.e. repurchase) from its owners at a date (call-date) and for a price (call-price) specified at the initial issuance.

It is possible to distribute funds proportionally among asset holders using the distribution function. This feature is like dividend payments, depending on their desired purpose. Distributions are paid in any distribution asset to everyone who holds the asset in proportion to how many units he holds. Distributions can be paid out to any assets that one owns and controls. One can freely select the currency in which distributions are to be paid out: BTC, XCP, or any other user-created asset.

Section 5.09   Antshares or NEO – a multi-use token

Antshares is a Chinese developed cryptocurrency. It is another cryptocurrency built on blockchain technology as a decentralized and distributed network protocol. It can be used to digitalize assets and accomplish financial business such as registration of assets, issuing certificates, making transactions, settlement, and payment through the P2P network. Its platform aims at the mainstream users. By sharing some commonality with Ethereum and Counterparty, Antshares intends to build a financial system bridging the real-world assets and cryptocurrency.[126] On June 22, 2017, Antshares developers rebranded the Antshares into NEO and converted Antshares coins (ANS) into Antcoins (ANC).

The ANS asset symbol became NEO in the 3rd quarter of 2017. So far, ANS is only traded on the Chinese exchanges and Brittrex. Like Ethereum, Antshares is a smart contract platform. Its developer is a Shanghai-based company founded in 2014, called Onchain. It runs the open-source project without owning it.

The similarity between Antshares and Counterparty ends at the asset digitization platform. The difference is more significant than the similarity. Antshares developers put particular emphasis on the compliance. Antshares has built-in KYC and AML APIs. Third-party payment providers, banks, and other financial institutions may utilize the Antshares protocol with compliance.

Currently, the system is compliant with rules set out by the Chinese Ministry of Industry and Information Technology (MIIT), as are 38 digital certificate companies in China that provide digital certificates of personal identity. When used in other countries, local rules may require local digital certifications.

Like Ethereum, Antshares runs smart contracts using the Mainnet client. Mainnet promises a much better ecosystem to issue and trade smart contracts. The Mainnet platform will also have some advanced personal identity features, allowing for instant authentication with existing banking infrastructure, much like digital certificate companies do today.

The first token to run on the open-source Antshares blockchain is the platform's Gas, a token named "Antcoins." The cap of Antcoins is set at 100 million Antcoins, generated over the next 22 years. Meanwhile, there is the same amount of Antshares, the token used for asset transfer. All Antshares tokens were created at once in the first block.

Antshares smart contracts are very simple contracts. They are for smart financial products and could be used for recording titles and assets like equities, creditor's claims, securities, financial contracts, credit points, bills, and currencies. They can also be

---

[126] https://neo.org/

Altcoins

used as equity crowdfunding, equity trading, employee stock ownership plans, peer-to-peer financing, loyalty programs, private equity funds, and supply-chain financing.

To prevent the DAO-like hack, the Antshares contracts are less interactive. The most important difference between Antshares and other cryptocurrencies is the use of delegated Byzantine Fault Tolerance (dBFT) discussed in Section 4.03 instead of PoW or PoS to secure its' blockchain. By using dBFT, the block verification is by consensus vote rather than mining. Other blockchain developers that use dBFT are IBM and Chain.com.

dBFT is a far more efficient way to achieve a similar security to Bitcoin's blockchain, without all of the cost of mining. The current version of dBFT used in the Antshares requires the consensus nodes to maintain a state table to record current consensus status.

So far, Antshares have two partners in China, Bitrees, and WeLand. Both are startups. Bitrees is the first Big Data analysis service of blockchain/cryptocurrencies in China. WeLand does the so-called Return Crowdfunding of physical products. Return crowdfunding is a new business model. It collects funds from the crowd to purchase physical products and promises to return the profit from the sales of these products to the investors.

In 2016, WeLand launched a "Landholder in Europe" crowdfunding on Taobao.com (not using Antshares), raising 4.1 million yuan from 19,000 participants to import European olive oil products and sold them in China, and returned the investment with profit to the participants. In WeLand's Return Crowdfunding last year, the average contribution from each participant is only $32, not a large amount. Most of them may not care about the outcome. However, for a larger amount of contribution, the contributor may want some kind of certificate or contribution agreement, such as share registration. However, it is difficult to do in the traditional form. With digital registration on the Antshares Blockchain, this problem can be solved with ease. By using Antshares platform, the future crowdfunding or Return Crowdfunding events will be able to implement digital registration of its return shares on the Antshares Blockchain. All participants will receive their copy of the smart contract.

Section 5.10   Bridging Bitcoin & EVM

Running smart contracts greatly enhance the capability of a blockchain. Suddenly, blockchain serves more than just creating and transacting cryptocurrencies. It finds all sorts of financial applications.

Ethereum is the de-facto leader of the smart contract system. However, PoW transaction validation imposes scalability limitation so that Ethereum is not feasible for most industrial applications. Ethereum is trying to solve the problem by gradually moving into PoW/PoS hybrid system.

Imagine if there is a way to rip all the benefits of private and trustworthy Bitcoin blockchain and running smart contracts like Ethereum and at the same time, without scaling limit. It will be a perfect match. Qtum - a Singapore-based Foundation developed such a technology.

Qtum is an open source, hybrid blockchain application platform. It builds an Account Abstraction Layer (AAL) to communicate between Bitcoin blockchain and Ethereum Virtual Machine (EVM). The AAL extends the Bitcoin Script language to transport code to the EVM and allows the EVM to operate within a UTXO environment, which simplifies the payment verification so that it can support mobile devices and IoT appliances.

Qtum merges the reliability of Bitcoin's blockchain with the smart contracts capability of Ethereum. In addition, the PoS consensus protocol allows Qtum applications much more scalable than Ethereum.

Qtum is compatible with pre-existing Ethereum smart contracts. Qtum platform can execute Ethereum smart contracts with little to no change to their code. Using Qtum, smart contracts and decentralized applications can run on a familiar platform with a robust environment. The target applications include mobile telecommunications, counterfeit protection, finance, industrial logistics, and manufacturing.

Altcoins

Qtum project takes the advantages the best of the Bitcoin and Ethereum platforms and opens wide varieties of applications not possible by Bitcoin and Ethereum. [127]

Section 5.11   Asset digitization

In June 2017, an organization called ACChain.org, operated by Guiyang Blockchain Financial Company in China, launched an ambitious blockchain project aiming to develop a tool and platform to digitize assets on a global scale. Such a digitization maps and segments all assets on a decentralized blockchain platform to realize tangible assets registration, issuing, and trading. It is essentially creating a unique digital copy of the asset in the blockchain platform. Such a digital copy of the asset can be tradable just as the real asset.

The trading of digitalized assets uses ACC (Asset Collection Coin). In the ACChain, each node can create a new block by copying the general ledger to form an asset chain. ACChain distinguishes itself from other popular cryptocurrencies in that ACC is emphasizing on commercial applications, assets digitization, and digital asset circulation, while Bitcoin is focusing on value transfer and Ethereum is focusing on smart contract. Therefore, ACC takes the position as a digital asset interchange medium to allow the exchange of all digital assets through ACChain.

There are three steps to generate a token for a digitized asset:

- It must meet certain criteria: such as durable, verifiable, traceable, etc.,
- Its published asset information must receive community recognition,
- It must receive approval from two-thirds of DAO.

On May, 9th, 2017, ACChain launched an ICO to raise 100 million units of ACC. 20% of the digital asset is used as incentives for members and 80% of the digital asset is locked in asset pool through a smart contract of the blockchain.

---

[127] https://qtum.org/en/

With this ICO, ACChain forms a DAO (decentralized autonomous organizations) to manage the fund from ICO. All members will manage capital; each transaction will require approval by 51% of the members.

The ACC ICO aims to structure supernode networks of digital assets around the globe and establish a digital currency of "Special Drawing Rights" (SDR). The SDR digital currency has not received the endorsement by IMF, the issuer of SDR. It is not clear how it will play out. It is quite possible that IMF will not allow ACChain to issue SDR digital currency since it has global implication way beyond any commercial application.

The intention of ACChain is to use SDR digital currency to spread the adoption of ACC. Each global node will establish regional "general ledger token" (GLT) for regional circulation, thus, ACChain envisions that the digital currency SDR will be the main exchange coin along with tokens of each international node's general ledger token in the international exchange.

In this ecosystem, each node's token can use GLT to realize regional circulation and each GLT and use ACC to realize international circulation. Those who control the international supernodes win the market. This is similar to today's role of dollars and national currencies.

ACChain launched a commercial real estate deal using blockchain technology in Dallas Fort Worth two weeks after its ICO in collaboration with Serene Country Home Group – a home developer in Texas. Serene Country Home Group launched the ICO of its own digital asset – RET (Real Estate Token) with the help of ACChain blockchain technology. RET will be a token to buy and sell properties in a new real estate project - Sendera Ranch in Fort Worth, Texas worth over US$600 million.

RET's use will not be limited to the trading of digital assets in the Sendera Ranch, it will also serve as the general token for the U.S. region and is for sale through ACChain's P2P sales supernodes. NPC, which stands for Node Primary Coin, is the coin issued by the supernode. NPC is a standard digital token based on ACChain.

BTC, ETH, and ACC are the exchange subjects for the NPC. There are three types of tokens: Equity token, application token, and commodity token. Different assets can have different tokens. Think NPC as the certificate of the ownership of asset and BTC, ETH and ACC are money. Money can exchange a certificate of the asset, and vice versa.

Puer is a special Tibetan tea, produced in Tibet and southwestern province of China, Yunnan. The high-quality Tibetan tea has stored by the third party, evaluated, authenticated, and right authenticated. The holders of the tea NPC can redeem the specified tea. Alternatively, they can also trade it for other tokens or exchange for other fiat currency. The NPC token itself is a proof of ownership of the tea. The price of NPC can rise or fall according to the market price of the underlying tea asset.

So far, ACChain has issued three different NPCs for three different qualities of the Puer tea via Fenghui International Finance in Shenzhen. They are Daguan Qianxue, Daguan Qiaomu, Daguan Jingong tokens. Likewise, Huayang, another company, has issued Jiangcoin for their product XiaoHongLiang Kunzi Pure Draft Liquor.

The growing number of institutions and companies joining ACChain to digitize their assets heralds a new stage in the blockchain technology: the emergence of commercial applications, which will be more useful to more consumers across the globe. Pencil Blockchain Company joined the ACChain DAO community to be one of the ACChain super nodes in North America for advertising ACChain and its commercial application implementation of the blockchain technology.

The company is in talks with a third-party fund management company to establish a mixed ETF Fund, the first building block of ACChain's financial services system. American Digital Asset Exchange Inc, a blockchain financial service institution, is also expected to join ACChain as a DAO member soon to establish a digital asset exchange transaction system for ACChain.

ACChain DAO offers different incentives to its members: the creation of block, asset issuing, and community voting. The ACChain DAO has encrypted PBFT algorithm to address the issue of abuse of right by the delegates, which makes the delegate's ability of accounting become more controllable. The ACChain system provides a command tool, it supports secondary chain development; each DApp corresponding to one secondary chain and the core logic of the secondary chain uses nodes to develop, the interface can adopt any front-end technology.

The Counterparty is another platform allowing the creation of assets in the blockchain. It opens a wide variety of applications. One can back Counterparty assets with tangible goods, such as gold. The so-called Turning complete smart contracts script can write programmable smart contracts and is one of the most powerful Counterparty features.

Users can write their own custom financial instruments and decentralized applications (DApp) using such a script. Smart contracts written in Counterparty script are compatible with Ethereum scripting, and all smart contracts can run on both platforms without code changes. Counterparty also allows users to develop and deploy Ethereum-style smart contracts on the Bitcoin blockchain.

When Counterparty assets are created within the Bitcoin blockchain, the assets actually exist in a different layer called Bitcoin block explorer. This is because the Bitcoin blockchain itself cannot hold any asset; therefore, an additional layer of block exploration is needed. There are two block explorers to store Counterparty assets in Bitcoin wallets: One is d the Blockscan. It allows one to view all the assets, broadcasts and transactions. The other is Counterparty Chain. It has features available to the public including Asset registration, transfer services, asset enhancement, asset vending machines, and enhanced broadcasting.

By doing so, Counterparty supports peer-to-peer asset exchange: users can trade assets with no intermediary and no counterparty risk. The platform upon which trading is done is Counterparty's decentralized exchange and the Bitcoin blockchain.

Altcoins

Chapter 6 Other blockchain platforms

Section 6.01 Permissioned vs. Permissionless

In Chapter 1, we have discussed briefly the Permissioned vs. Permissionless blockchains. In the blockchains like Bitcoin and Ethereum, any user can join the network and start mining. One does not need anyone's permission to join the blockchain. In addition, one does not have to prove his/ her identity to anyone. As long as one can commit the processing power to be part of the network and extend the blockchain. The permissionless blockchain is like a public park that anyone can walk in. Since the network is open to all, the only way to make the transactions secure is that the participating nodes do the block verification by performing PoW/ PoS/ dBFT type of work.

As more large corporations are looking into the blockchain technology to explore its potential use, it occurs that applications can be very specific and only involved parties need to participate. For example, if a bank develops a blockchain application to serve its clients, naturally, it will only allow its clients to use such application. In this type of Permissioned blockchain, the central authority establishes the trust. To open an account at a bank or apply for a credit card, the necessary ID and financial information of the applicant already serve as a "trust".

There are two aspects of the permission: one is the right to access, the other is the right to validate transactions. In the Permissioned blockchain, only a restricted group of users by invitation has the rights to validate the block transactions. The right to validate transactions is more restrictive than the right to access. In the real world, there are also different levels of

Other blockchain platforms

privilege: a bank's client has the right to see his account balance and to write a check, but he does not have the right to change the account maintenance fee or to see other clients' balance.

Similarly, in a blockchain belonging to a bank, the right of validation is restricted to the bank's network notes, and the right of access is restricted to bank's clients. Therefore, the Permissioned blockchain is like a private club where only admitted members can enter. In addition, there are different levels of members that have different privileges and rights.

The Permissioned blockchain has an owner, who approves the participation membership. The owner builds the blockchain to serve his purpose. When banks and financial institutions build blockchain applications, they use the Permissioned blockchain to do so. National digital currency also uses the Permissioned blockchain because the digital currency has an owner – the central bank.

The Permissionless blockchain needs to provide coins as the incentive to validate the blocks and built the blockchains, which requires resources. On the other hand, in the Permissioned blockchain, token does not serve this purpose anymore. Unless transactions require tokens, the blockchain does not need tokens. Some Permissioned blockchains are token-less blockchains. The owner controls the validators and block-creators; they do their job for different reasons other than receiving tokens. To operate a blockchain application is a part of their business operation.

As a result, the validation work can be very different in these two types of the blockchain. In the Permissionless blockchain, the use PoW mining or PoS minting is necessary to achieve consensus. Bitcoin blockchain uses hashing power to build trust. Ethereum currently is using PoW but may adopt both PoW and PoS as we have discussed.

Permissioned blockchains already operate based on the trust of the institution who owns it; therefore, their blockchains do not require PoW to validate transactions. The institution (e.g. bank) provides the trust. The most important attribute of their blockchain application is security and efficiency. They can use a simplified version of PoS or dBFT, or use very different kind of consensus algorithms like RAFT, Paxos or PBFT.

Even without the PoW or PoS consensus mechanism, the Permissioned blockchain still has the following advantages:

- Privacy – only members have rights to view the transactions.
- Scalability – by not using the resource-intensive PoW, a Permissioned blockchain can be easily scaled up.
- Access Control – a Permissioned blockchain can restrict access to the data within the ledger as the owner (bank) desires.

SETL, [128] a startup based in London, has created a Permission-based blockchain settlement and payment system, that can move cash and assets in real-time to settle market transactions. It maintains a distributed ledger of ownership and transaction records, simplifying the process of matching, settlement, custody, registration and transaction reporting. The platform enables investors and distributors to easily subscribe and redeem fund units via a direct connection with the Asset Management Company, thereby removing the need for the transfer agent, which in turn reduces transaction costs. The platform increases transparency, optimizes operational workflow and enables the development of new value-added services.

In summary, the Permissioned blockchain is a closed, private blockchain with an owner. It has a certain degree of centralization. In a way, they are similar to the comparison of internet and intranet. The Internet is a robust high capacity global network for transmitting information within an organization. This enabled many enterprises to make use of virtual private networks (VPNs), which use the Internet as a backbone but encrypt the organization's traffic over these public pipes. VPNs allow enterprises to enjoy the Internet's economy of scale while ensuring that their data is not visible to outside observers. One can imagine a comparable process playing out between the permissionless/ public blockchain and Permissioned/ private blockchains.

Section 6.02   Identity, Transaction and Content MDL's

---

[128] https://setl.io/

Other blockchain platforms

There are three types of Permissioned or private MDLs: Identity MDLs, Transaction MDLs, and Content MDLs. As the name implies, the Identity MDLs holds the identity of the entity, which has access to a Transaction MDL. A Transaction MDL holds the hashes of all transactions. The Content MDL holds all of the original documents.

The Content MDL can be very large, like the database in a data center; nevertheless, it is still distributed. Since all documents in the Content MDL are hashed and stored in the Transaction MDL, any alternation in the original document in the Content MDL will have a discrepancy with its hash in the Transaction MDL. Therefore, the Content MDL is safe from unauthorized alteration.

One can combine these three types of MDLs into a single MDL. The three-in-one MDL has the advantages of simple in the data structure, easy to implement and to search, reduced risk of data loss, the simple link between data and content, more options for storing sensitive content. The disadvantages are that the MDL will be bigger and there is a potential lack of oversight over sensitive personal information. Meanwhile, sharing with other MDL's can be more difficult.

The only structural difference between the Identity, Transaction MDLs, and the Content MDLs is that the first two MDLs have a fixed-length hash field while the Content MDL has a variable length field to hold documents, pictures, videos or spreadsheets, and other types of documents.

When these three types of MDLs are separate MDLs, the ability to link MDLs is important, so that their storage is segregated and yet their functionality is integrated.

Separate MDL's make the data sharing much more flexible. One MDL can interface with many different MDL's. For example, the MDLs of different organizations may want to share the same Identity MDL. Or the police department may want to use the same Identity MDL as the department of motor vehicles. The insurance company may want to use the same Identity MDL as the hospitals.

The economics of multi-users is driving the modularization of MDLs, which leads to the cross-organization systems. The flexibility makes it possible to share part of the MDL but not all of the MDL, using Quorum-like public/ private states, or even multiple private states in the database. For example, a hospital may share part of its private identity MDL with one insurance company and a different part of another insurance company. Likewise, because each insurance company has its own provider network, an insurance company may share part of Identity MDL with one provider and a different part of another provider.

With all these MDL's established in a cloud, the transactions between different organizations can be fast and secure and only relevant parties are involved. Eventually, the cross-organizational MDLs can be so extensive that it would link most of the organizations together.

In order to integrate several blockchains into one, standard and data sharing protocols need to be developed. A standard protocol involves registration, inspection, certification, and checking of the underlying MDL's. The data sharing protocols are to validate data links to different chains networks. This opens up completely new possibilities for different MDL's to work together when necessary.

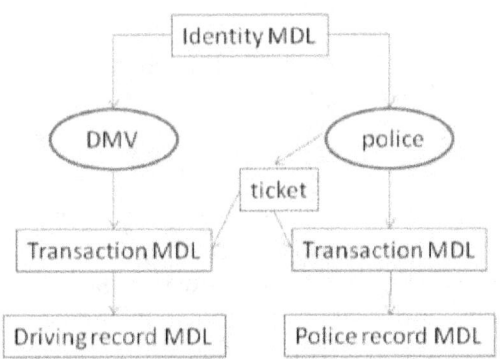

Figure 6-1 Example of a shared MDL

It is essential to develop the ability of these three functional different MDLs to exchange information even they do not belong to the same entity, or they have to be modular. An example Figure 6-1 shows two organizations – DMV and Police

Other blockchain platforms

share the same Identity MDL. However, each one of them has different Transaction MDLs and Content MDLs.

When a police write a traffic ticket, he creates a transaction to the transaction MDLs, and they become a permanent ledger entry in both the DMV and Police Driving record MDL and Police record MDL. These links are not necessarily to be in the same cloud. The P2P network serves as a bridge between clouds. The integrated MDLs of identity, transaction, and content MDL's form an archiving system as a ledger itself. They can also support/interact with other ledgers.

Multiple databanks, such as healthcare data, insurance certificates, credit history, contracts, real estate deeds and many other documents can be classified, encrypted and tracked. This MDL can be distributed across several platforms, over the internet, or in the cloud, and provides the same level of security and the convenience of asset transfer.

What we have described is a concept called "sidechaining". [129] Sidechain is a separate blockchain from the main blockchain, which has tokens. Sidechaining is the mechanism that allows the sidechain to use tokens from main blockchain securely. The tokens can always move back to the original chain if necessary.

For example, a private Ethereum-based network that had a linkage allowing Ether to move securely from the public Ethereum main chain onto it and back would be considered a sidechain of the public network.

However, the idea of sidechain extends to more than just tokens; it can exchange anything of value cross MDLs as well. In doing so, suddenly MDL is no longer isolated. Valuable assets can move across different MDLs. Once the modularity of the MDL is developed, a true internet of value will be a reality.

The concept of immutable data could materially alter the way society views identity, privacy, and security. MDL

---

[129] "A simple explanation of Bitcoin sidechain": Richard G. Brown, https://gendal.me/2014/10/26/a-simple- explanation-of-Bitcoin-sidechains/

technology may in future bring a network of the omnipresent database, but before MDL technology solutions are more widely adopted, a number of conditions need to be met: developing standards for interoperability; and creating a cohesive regulatory position on MDL technology applications, among other things.

Section 6.03   Identity MDL

In the sharing economy, the identity becomes the most important trust element. For example, in the bike-sharing platform, without identity, anyone can just take the bike and never return it, or get a free ride without paying for it. Therefore, identity is necessary to do transactions and hold assets.

Identity MDLs could make possible a wide array of asset related tasks. Efficient identity systems are fundamental to efficient financial and trading systems. The persistence and pervasiveness of MDL's make them ideal for providing a lifetime record.

Theoretically, a single Identity MDL can serve any number of transactions that require an identity. Thus, the idea is born to develop Identity MDL as a stand-alone MDL that can serve many different clients.

In June 2017, a Decentralized Identity Foundation (DIF) was launched at the Consensus blockchain tech conference in New York.[130] The founding members include Microsoft, Accenture, IBM and many others.

The goal of the DIF is to develop a system using blockchain technology to register self-sovereign identifiers that no provider owns or controls. It will be able to lookup and discover identifiers and data across decentralized systems, and for users to store sensitive identity data securely, and to control what to share with others.

Blockstack,[131] one of the foundation's founding members, has released a set of programming tools to create a decentralized

---

[130] http://identity.foundation/
[131] https://blockstack.org/

Other blockchain platforms

internet, and to construct Identity MDL. Blockstack toolset decentralizes the application layer of the internet by enabling decentralized storage and authentication. Users run DApps through the Blockstack browser. Information is encrypted and stored on users' personal devices.

Microsoft has collaborated with Blockstack Labs and ConsenSys to build an open-source identity platform aimed at integrating the Bitcoin and Ethereum blockchains. Blockstack collaboration with Microsoft and ConsenSys aims at a global standard for blockchain identity applications.

Microsoft will launch an open source framework with the cross-chain applicability on Azure, where developers can build their own identity applications. Microsoft's Identity MDL will connect with the Ethereum blockchain via ConsenSys' uPort solution [132], and with the Bitcoin blockchain via Blockstack's OneName.[133]

The project finds growing interest among global organizations in the use of blockchain technology to address long-standing identity issues in many nations. A readily available Identity MDL, such as the one built by Microsoft, can facilitate the creation of new Identity MDLs by not having to build its own identity MDL from scratch.

DIF is not alone in developing Identity MDL. The open-source Hyperledger Project, as we discussed in Section 6.11, is also working on a decentralized identity project called Indy.[134,135]

Identity MDL could empower people with personal data storage and management, permission frameworks for access by third parties such as banks or insurance companies, and even distributed reputation ratings. Such applications could reduce

---

[132] https://www.uport.me/
[133] https://onename.com/
[134] "Hyperledger welcomes Indy", https://www.hyperledger.org/blog /2017/05/02/hyperledger-welcomes-project-indy
[135] "Next consensus architecture proposal." E. Androulaki, et. al. https://github.com / hyperledger/ fabric/ blob/master/proposals/ r1/Next-Consensus-Architecture-Proposal.md, 2016.

identity and fraud, increase confidence in products and lower rates thus increasing coverage.

There is an infinite number of potential applications. For example, it can link to an education record, driving record, tax record, medical record, marriage record, employment record and many more. MDL technology and related applications could transform the way we manage digital identity, personal information, and history.

Not only individuals have an identity, but also a juridical entity, such as a company. DueDil [136] offers a platform that provides authoritative data over 40 million private companies in Europe so that the information available on public companies now for the first time will be available on the private companies as well. This helps businesses to evaluate risks dealing with these private companies.

An ID scheme relying on decentralized MDLs combining a public ledger of records with an adequate level of privacy could rival state-backed identity systems. One day it may replace the driver license, social security card, and passport. Eventually, one's fingerprints, face, and any other biometrics, or even DNA will be in the Identity MDL.

To establish an Identity MDL, the first step is to input identity data onto the ledger. These data include the data in driver license, passports and any form of ID cards, birth or death certificates, signature records, criminal records, educational degrees, professional qualifications, certifications, human resources records, medical records, bank records, business transaction records, location data, genome and DNA, genealogy trees, etc.

The data entry can be from any system using identity validators. Data from different sources are cross-referenced. Any data belonging to a person can be relevant, such as vital record, education record, health record, financial records, etc. Data will be collected since birth and continues to be collected throughout one's lifetime. Combining authentication and personal data

---

[136] https://www.duedil.com/

Other blockchain platforms

management functionalities with secure MDLs could lead to new frameworks for identity management.

Any activity of an individual, when recorded in a transaction MDL, becomes a transaction record in the MDL. Each of these transactions is a permanent record to an identity in the Identity MDL. When multiple MDL's are linked, the dots become a line and a picture of the identity emerges.

MDL technology and related applications could transform the way we manage digital identity (ID), personal information and history. An Identity MDLs could surpass many of today's state-owned identity systems such as passport, driver license etc, in terms of the available information, speed of verification, fraud prevention, and all-in-one feature. A number of digital ID schemes are emerging, including OpenID Connect,[137] which allows clients to request and receive information about the authenticated session.

India is embarking on a comprehensive national ID system. We will discuss India's Aadhaar system in Section 6.04. India is not alone. Governments around the world are trying to set up digital ID systems and authentication processes. Countries like U.K., Estonia and many others all have the similar projects in progress. The U.K., for example, unveiled Gov.UK Verify[138] in September 2014, a proposed public services identity assurance program that might use a network of trusted and vetted third-party providers instead of relying on a centralized database. Estonia has been operating a national digital ID scheme for a decade and is now updating to MDL structures.

Since an entity such as government owns an Identity MDL, it is a Permissioned Identity MDL. By possessing such amount of personal data, the government holds tremendous power over its people. This has raised fear for the government creating such an Identity MDL. Therefore, creating a trusted and widespread digital ID system is more than a technical task. It is also a social task. The nationwide implementation of such a system may meet much resistance, as it becomes a tool of absolute

---

[137] https://www.pingidentity.com/en
[138] https://www.gov.uk/government/publications/introducing-govuk-verify/introducing-govuk-verify

totalitarian control with mistrust. Establishing an efficient identity system is the core global development challenge for MDLs.

Like any system, there are pros and cons of a nationwide Identity MDL. On the positive side, an efficient identity system is fundamental to efficient financial and trading systems. With the explosive growth of MDL into the internet of value and the storage and transfer of assets, such as land and property, business ownership, regulatory records, court records, smart contracts, vital documents, medical and health records, or eventually genome and DNA records, it is unimaginable without a proper Identity MDL, such a Transaction and Content MDLs can function securely.

In contrast to the Permissioned Identity MDL created and owned by the government, Permissionless Identity MDL belongs to the public. It retains all the advantages of the Permissioned Identity MDL but without a controlling authority.

In the Permissionless Identity MDL, one has the access and control over one's own data. This scheme will empower individuals to store, update and manage access to their identity data much like today's social media. However, social media networks fail to meet basic trust requirements as most data having no verification, and the data in social media are not secure.

The Permissionless Identity MDL will require the data to be authenticated and notarized by identity validators, which serve as a "notary public" of data. An identity validator might be a government agency, an accounting firm or a credit-referencing agency. People go to an identity validator to encode biometrics, e.g., DNA, retinal scan, photo, facial scan, finger vein identification, thus time-stamping physical identity. Validators notarize the data as a trusted third party. In InterChainZ, the identity validator is a notary of data onto a personal or corporate MDL. Once validated, the data in the Permissionless Identity MDL can be trusted, and is secure, private and safe from any misuse.

Co-stamping and validation themselves are important aspects of development for the Permissionless MDL. There are huge varieties of identity data. For an individual, this could be health record, driving record, financial record, biometric data,

Other blockchain platforms

academic record, etc. The validation or co-stamping of these data can be complex.

How to develop an efficient and functional Permissionless MDL free from government or any other authority control is the key to maximize the benefits and minimize the downside. Combining authentication and personal data management functionalities with secure MDLs could lead to new frameworks for identity management. If successful, such identity schemes could remove government monopolies in managing their citizens' identities and data. At a time where access and control over one's own data are becoming increasingly sensitive, empowering individuals to store, update and manage access to their data seems rather appealing.

Section 6.04   Digital currencies

Digital currency provides an alternative to traditional fiat currency as a store and transfer of value. Bitcoin is the most popular digital currency today. Yet, as we have discussed in the previous chapters that Bitcoin has its limitations.

In addition, country's government wants to maintain the absolute control of issuance, circulation, and supply of its currency. Many countries, including Russia, UK, China, India, Netherland, Ecuador, Sweden, Denmark and more are contemplating to introduce national digital currencies. Bank of England envisions using cryptocurrencies to settle interbank transactions directly between parties without an intermediary. In Africa, Senegal has already announced a blockchain-based national digital currency.

In this section, we will look into the Indian system in more detail here since it is in a more advanced stage and may have an implication for more than 1.3 billion population. Its success (or failure) can be an interesting experience for the rest of the world.

In November 2016, the Indian government announced the demonetization of all 500 Rupees (US$8) and 1,000 (US$16) banknotes.[139] The government claimed that the action would curtail the shadow economy and crack down on the use of illicit and counterfeit cash to fund illegal activity and terrorism. The sudden nature of the announcement—and the prolonged cash shortages in the weeks that followed—created a significant disruption in the economy, threatening economic output. At the same time, India introduced India Stack as a part of Digital India program.

The Digital India program is more than just a digital currency program. Through a program called Aadhaar, it digitizes the identification of all its citizens. It is much like to digitize the social security number/ driver license/ passport for all the citizens in the US. Once this system is fully implemented, every Indian citizen is immediately identifiable through his or her records in the national database.

The government has launched an open Application Programming Interface platform, which includes the Aadhaar for authentication, e-KYC for customer interface, and e-Sign. India Stack's API allows governments, businesses, startups, and developers to utilize a unique digital Infrastructure to facilitate India's paperless and cashless service delivery.

eKYC is a paperless Know Your Customer (KYC) process, where a person's Identity and address are verified electronically through Aadhaar Authentication. It replaces the current processes using physical photocopies of the original documents for ID proof and Address proof.

eSign service allows an Aadhaar ID holder (eventually every Indian citizen) to electronically sign a document anytime, anywhere, and on any device legally in India. It facilitates a significant reduction in paper handling costs, improves efficiency, and offers convenience to customers. The e-Sign service is carried

---

[139] "India tried to get the 'black money' out of its banking system — it ended up doing the opposite", CNBC, https://www.cnbc.com/2017/09/07/demonetization-reserve-bank-of-india-suggests-that-demonetisation-allowed-black-money-to-enter-banking-system.html

Other blockchain platforms

out on a backend server of the e-Sign providers, which are trusted third party service providers called Certifying Authorities (CA).

A Unified Payments Interface is on top of India's Immediate Payment System. Tech start-ups can use this framework to develop mobile apps and make services available to a large section of the population.

A system called UIDAI (Unique Identification Authority of India), [140] independently from the digital currency is a Demographic Authentication and Biometric Authentication system in India. It authenticates a person's identity against the data in the demographic database, which eventually will contain the data for all billions of Indian citizens. During the authentication transaction, the resident's record is selected using the Aadhaar Number (ID number) and then the demographic/biometric inputs are matched against the stored data provided by the resident during enrolment. Aadhaar is a 12 digit random number, issued to the residents of India, by the Unique Identification Authority of India (UIDAI).

Every resident will receive an assigned Aadhaar number by submitting his/her minimal demographic and biometric information during the enrolment stage, or when a baby is born after the system is implemented. The biometric data include irises scan, fingerprints of the ten fingers. Without Aadhaar or the UIDAI issued ID number, one will not be able to use digital cash, which will be the only form of cash in India. The Aadhaar program is already at such an advanced stage that over billion Indians have obtained their Aadhaar number. And a new mobile phone network from the Reliance Industries Ltd. conglomerate recently used the Aadhaar network to enroll 100 million subscribers in three months.

The government now can transfer welfare payments directly to Aadhaar-linked bank accounts, cutting out India's notoriously corrupt intermediaries. Today, many Indians do not even have birth certificates. Once this system is in place, anyone can be identified through a centralized achieve, which stores birth, education, marriage, and many other identity-related data in one

---

[140] Unique Identification Authority of India website"
https://uidai.gov.in/

central database. Aadhaar is an ideal platform for the digital currency verification.

The State Bank of India (SBI) launched blockchain-based know-your-customer (KYC) processes in 2017. A consortium of 27 banks develops the KYC system. The platform is called "Bankchain", which is a system for sharing information about customers. [141] The consortium is working with Primechain Technologies, a Mumbai-based startup, as well as IBM and Microsoft. The objective is to reduce fraud and an alternative system for remittances and SWIFT.

SBI has also planned to explore new technologies like blockchain, artificial intelligence, machine learning, among others. In such a system, all the transactions and life events are handled by blockchain. These events can be birth, graduation, marriage, transaction of real estate and many others.

However, instead of using the anonymous cryptocurrency addresses, the system will use the real identity of the person, including all his biometric data. Once the system is in use, nothing an individual does can be hidden from the government, as long as it involves a transaction even as small as a petty cash payment.

This Aadhaar project is in parallel with the creation of the India stack, the Indian digital currency. It will be truly revolutionary that when implemented, even a petty cash transaction, such as buying a cup of coffee, will be carried out electronically and the identity of both payer and receiver will be authenticated by the UIA. The world's billions of people will be instantly identifiable one day in the near future with everything he or she has done since birth to everything he or she possesses. The power of Identity MDLs will transform the world forever.

If other governments follow the Indian model, the day of privacy will be numbered. The identity data will be used to authenticate the person when he or she makes a purchase or receives a payment, no matter how small it is, or makes a smart

---

[141] "India's 'BankChain' Consortium Launches Blockchain KYC System", Stan Higgins, https://www.coindesk.com/indias-bankchain-consortium-launches-blockchain-kyc-system/

Other blockchain platforms

contract transaction in the same platform. The big brother society will no longer be a fiction.

Revelation 13:17 in the Bible entitled "Mark of the beast" says: "And the second beast required all people small and great, rich and poor, free and slave, to receive a mark on their right hand or on their forehead, so that no one could buy or sell unless he had the mark the name of the best or the number of its name."

The trend is universal, more central banks and governments are working on digitizing their national currencies. China and Russia are investigating the potential of Ethereum as the base protocol for a digital RMB and Ruble respectively. Currently, the Royal Chinese Mint, the subordinate unit of China Banknote Printing and Minting, is testing Ethereum and its ERC 20 token standard to digitize the RMB. The Royal Chinese Mint is at the forefront of research and exploration into digital money. By utilizing the ERC 20 token standard, the Royal Chinese Mint is considering the possibility of releasing unique tokens that are compatible with the Ethereum network. An Ethereum-compatible token would grant higher liquidity and interoperability. The Royal Chinese Mint is actively promoting the application of blockchain technology in finance and related fields.

Russia is also looking into Ethereum and its potential in the finance sector. In 2016, the Bank of Russia announced the development of an Ethereum-based interbank blockchain prototype called Masterchain. Some of the largest commercial banks in Russia participated in the pilot test.

How these significant changes will affect companies, such as PayPal and highlights the advantages of using third-party payment services over the credit card payment methods.

Section 6.05   Token-less MDLs

We have discussed that token is not necessary for a Permissioned blockchain. The token is a medium of the incentive to do the validation work or a means of exchange of value. When the tokenless blockchain was first proposed, there were some doubts whether there were benefits to MDLs without PoW/ PoS validation mechanisms, which are the core of MDL.

In a Permissioned blockchain, there is an owner of the blockchain. All the network nodes belong to the owner. Permissioned ledgers normally have strong structures for multiple parties, e.g., regulators or the ledger is within a single organization. In this type of environment, the institution of the blockchain provides the trust.

In fact, to make the transaction securely processed is the strong enough incentive. Tokens are not necessary to provide incentives for the nodes to do the validation work. At the same time, the smart contract governs the terms and conditions of the exchange of value. Therefore, in the Permissioned blockchain, tokens are not necessary.

InterChainZ Consortium, launched by Z/Yen, a commercial think tank based in London, proposed MDL applications without using currencies or tokens.[142] By getting rid of the token, the transaction speed is improved. InterChainZ Permissioned ledgers can achieve 5,000 transactions per second, which is several hundred times faster than the transaction times of token-based MDL's.

The project developed a number of MDLs that directly stored documents, as well as MDLs that only recorded the "hash" of documents. InterChainZ demonstrated how MDL technology might provide such capabilities for current financial services. It builds software providing an interface to MDLs for tasks performing a sharing economy. It has demonstrated MDL's functionality in an identity validation service, and the functionality to use MDL's to validate company's identity and report on their finances.

Section 6.06   BigchainDB

BigchainDB platform, developed by a German company called BigchainDB GmbH in 2016, is an open source system that adds blockchain-like features to a distributed database.[143] Its objective is to build an MDL with both the advantages of Bitcoin-

---

[142] www.zyen.com
[143] https://www.bigchaindb.com/

Other blockchain platforms

like blockchain and the traditional distributed database. It is taking an opposite approach as most of the MDL's. Instead of building the database in a form of a distributed ledger, it builds a distributed ledger out of a database.[144]

BigchainDB takes advantages of distributed database's own built-in consensus algorithm to tolerate benign faults, and use the DB's built-in communication, for great savings in complexity and for reduced security risk. Like any MDL, it has decentralized control, immutability and can transfer digital assets.

Comparing to the Bitcoin blockchain, it boosts much faster transaction speed with sub-second latency and can store much more of data. It removed all the elements in the Bitcoin-like blockchain that slow down the operation and limit the scalability. For example, it uses a hash chain instead of the Merkle tree, and PoW was replaced by consensus. Mechanisms, such as the shared replication, the reversion of disallowed updates or deletes, regular database backups, cryptographic signing of all transactions, and blocks & votes guarantee the immutability / tamper-resistance. Each vote on a block also includes the hash of a previous block. Any entity with asset-issuance permissions can issue an asset. A new owner can acquire an asset if they fulfill its cryptographic conditions.

The Permissioned blockchain supports custom assets, transactions, permissions, and transparency. BigchainDB uses Tendermint for consensus and transaction replication. Tendermint is a version of the Byzantine Fault Tolerant protocol. Tendermint can replicate an application on many machines securely and consistently.

Tendermint consists of two components: a blockchain consensus engine and a generic application interface. The consensus engine, called Tendermint Core, ensures that the same transactions are recorded on every machine in the same order. The application interface, called the Application BlockChain Interface (ABCI), enables to process the transactions in any programming language. Tendermint is easy-to-use, simple-to-understand, and useful for a wide variety of distributed applications.

---

[144] "A BigchainDB primer", https://www.bigchaindb.com/

BigchainDB also gets rid of the token. It can also be configured as Permissioned or Permissionless MDL. Increasing numbers of nodes can increase its throughput and capacity.

An existing database can add BigchainDB to gain decentralization benefits. One can also build a BigchainDB database from scratch into a full-blown decentralization ecosystem. Since it is a database, it can store documents, contracts or any form of data just like traditional database without the scalability issue.

In a Bitcoin-like blockchain, every node has a replica of the blockchain. To do so for the content storage, the database can be huge. It is impossible that the participating nodes have the capacity to store such big database. Therefore, it is unlikely to replicate blockchain at every node. The solution is the partial replication. For example, in Netflix database, each data has only three replications. Partial replication protects the data and yet does not overload the storage capacity. As the database grows, the number of nodes grows proportionally, but the number of replications remains the same. Thus, the total storage capacity increases with the number of nodes. Most modern distributed DBs have a linear increase in capacity with the number of nodes. In such a system, database's latency and network usage stay the same. Therefore, it scales linearly.

BigchainDB takes the same approach of partial replication to limit the blockchain size. In summary, BigchainDB is MDL built from a traditional database by adding blockchain characteristics including decentralization, immutability and built-in support for creation & transfer of assets.

Section 6.07 Corda

Corda is an open-source distributed ledger platform.[145] It is a product of R3, founded in 2014 as a distributed database technology company, and headquartered in New York. R3 considers Corda a distributed database technology rather than an MDL technology.

---

[145] https://www.corda.net/

Other blockchain platforms

Corda aims for use with regulated financial institutions. Its approach is actually quite appealing to the established financial institutions.[146] It leads a consortium of more than 70 of the world's biggest financial institutions. This represents the biggest joint effort among banks, insurers, fund managers and other financial institutions working on applications using blockchain technology in the financial markets.

In a traditional distributed database system, the multiple nodes that share the distributed database belong to a single entity or owner, such as a company. There is already a trust between the distributed database system as a whole and its users. Each node in the system trusts the data that it receives from its peers and nodes are trusted to look after the data they have received from their peers. The only validation necessary is when this distributed database exchanges information with the outside world.

Corda's distributed database does not reside in a system with the single owner; rather, it can spread to many independent and unrelated nodes on the internet. In order to do so, Corda must build the trust by independently verifying data nodes receive from each other.

Corda uses the cryptographic techniques, the key element in the blockchain technology, for the security of the shared records. Even though Corda does not use PoW and mining, it has many qualities of Bitcoin: immutable states, transactions with multiple inputs and outputs.

Nevertheless, Corda does not use blockchain and broadcast the transactions to all the nodes in the network. Instead, Corda sends data only to the relevant parties. Corda relies heavily on secure cryptographic hashes to identify parties and data.

In a Corda system, the ledger has a different definition: it is not a list of all transactions but is a set of immutable state objects. It defines states and transactions, where every transaction consumes existing states and produces a new state (Figure 6-2). The issuer and affected nodes must endorse transactions to ensure

---

[146] "Corda: an introduction": Richard G. Brown et. al. https://docs.corda.net/_static/corda-introductory- whitepaper.pdf

the correctness according to the underlying smart-contract logic governing the state to be valid.

Only nodes affected by the transaction receive the new state. In other words, only the parties, having the privilege to see the data, can share the data. This improves the privacy, scalability, legal-system compatibility of a traditional distributed database. At the same time, it avoids the broadcast of transactions to unrelated parties as in the case of other MDL's. This provides means for partitioning the data among the nodes.

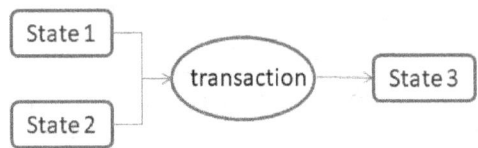

Figure 6-2 Corda transaction consumes existing states and produces new state.

Notary, which consists of independent parties running the same contract code and validation logic, verifies the validity and uniqueness of the transaction to all its input states. It must ensure that each state can only be consumed once. The notary also has the responsibility to timestamp the transaction including the states pointing to it. The state may designate an asset represented by the network, which can be anything controlled by a smart contract or tokens.

An immutable state object in Corda system is the hash of a legal prose and contract code that governs its transitions. Corda uses smart contract, timestamping, and a framework, which simplifies the process to achieve the immutable state objects. A transaction produces a new state, which is the change in the input state object.

Corda supports smart contracts with many unique features. For example, it records and manages financial agreements or any other shared data between parties in the legal constructs and regulation by including regulatory and supervisory observer nodes. It organizes workflow, supports consensus, and validates transactions between parties to the transaction.

Other blockchain platforms

In summary, Corda system is like a hybrid of a mutual distributed ledger and distributed database. Corda application is known as CorDapp (short for Corda Distributed Applications). [147]

Section 6.08 HydraChain

HydraChain[148] is an open source, Permissioned Distributed Ledger platform based on Ethereum. In essence, it is the Permissioned version of Permissionless Ethereum. Its protocol is fully compatible to Ethereum. HydraChain supports for creating a permissioned distributed ledger using the Ethereum infrastructure.

Like BigchainDB, HydraChain uses a proprietary consensus protocol derived from Tendermint – a consensus based on the Byzantine Fault Tolerant consensus protocol. The validators negotiate new blocks. A quorum of the validators signs off the new blocks before adding them to the chain. Therefore, there is no fork or revert. The new blocks are created only when there are pending transactions. The block generation is very fast, in less than a second.

The important feature of HydraChain is that it enables to develop smart contracts in the Python high-level language, which is easy to use and debug. The contract written in Python in HydraChain is compatible with EVM based contracts written in the Solidity or Serpent languages and can co-exist on the same chain. Therefore, it is fully Compatibility with the Ethereum Protocol. The code runs significantly faster than on the codes in Solidity or Serpent.

Users can also customize Hydrachain to their needs. This makes Hydrachain highly useful to many applications.

Section 6.09 MultiChain

MultiChain blockchain platform[149,150], launched by Coin Sciences, allows financial institutions to create public or private

---

[147] https://docs.corda.net/cordapp-overview.html
[148] https://github.com/HydraChain
[149] https://www.MultiChain.com/

blockchains. MultiChain is a Permissioned blockchain, and an open-source blockchain platform, based on Bitcoin's blockchain, for multi-asset financial transactions. It uses Byzantine Fault Tolerant variation of consensus protocol. The user can easily tailor these blockchains to his particular needs and deploy either within or between organizations. It facilitates the institutional financial entities to deploy blockchain technology. It also provides the privacy and control required. [151]

The name, MultiChain, comes because users can connect to the number of blockchains they created using MultiChain. The owner can also issue assets on the blockchain it created, track and verify assets at the network level, perform multi-asset and multi-party atomic exchange transactions, which meet the requirement of multi-asset and multi-party simultaneously. It can also create multiple key-value, time series or identity databases on a blockchain for data sharing, timestamping and encrypted archiving. [152]

MultiChain has the flexibility to allow customers to control many aspects of the blockchain. The chain can be private or public. The user can control the target time for blocks, limit the visibility of the ledger to certain participants, and set the controls on transactions permitted. The user can also decide who can connect to the network; how these entities interact; and the maximum block size and metadata to be included in transactions, among other features.

Because MultiChain uses Bitcoin's protocol, transaction, and blockchain architecture, an application of MultiChain can move between Bitcoin and MultiChain with minimal changes to its code. It, therefore, provides smooth transitions between private blockchains and the Bitcoin blockchain back and forth. Its

---

[150] MultiChain whitepaper: http://www.MultiChain.com/white-paper/

[151] "An open source project for creating a private blockchain ecosystem preloaded with MultiChain and related tools", YobiChain, https://github.com/Primechain/yobichain

[152] MultiChain official site: http://www.MultiChain.com/developers/

Other blockchain platforms

interface is fully compatible with that of Bitcoin Core. It can also act as a node on the Bitcoin network.

Like other permissioned blockchains, only nodes in the permitted list can validate blocks and participate in the protocol. Their public keys can identify them. The miner of the first "Genesis" block automatically receives all privileges, including administrator rights to manage the privileges of other users. This administrator grants privileges to other users in transactions whose outputs contain those users' addresses together with metadata denoting the privileges conferred. However, to remove the privileges of a user requires voting of the existing administrators. The voting process increases transparency of the network.

As the MultiChain platform evolves from Bitcoin, its consensus mechanism is also "mining". However, in the permissioned model, the nodes do not solve computational puzzles. They generate blocks by the administrators subject to certain constraints. If two different nodes generate a valid block at the same time, and any other node will append the one of which it hears first to its chain. This creates a fork. The longer branch will win, and all nodes will continue mining on top of that. Transactions in the blocks of the shorter branch re-enter the memory pool of nodes. That means that they will start the confirmation process over again like any other new transactions.

The modifications to privileges propagate quickly to all nodes in the network. To avoid that one participant can monopolize the mining process; MultiChain implemented a restriction the number of blocks that the same miner can create within a given window.

The user can also grant permissions on a limited basis or to a fixed range of blockchain numbers. In MultiChain, its administrators have the authority to set a list of permitted users that can act as nodes. The user can also grant privileges using transactions with special metadata.

It uses consensus mechanism of the Practical Byzantine Fault Tolerance with one validator per block. MultiChain validation process enables miners to approve transactions in a random rotation. A MultiChain user can also set the percentage of the miners' approval needed to record a block.

MultiChain also allows the chain's owner to control the maximum block size, so that scaling is not a problem. MultiChain uses the public key and private key to restrict blockchain access to a list of permitted users.

The scope of the permissions extends to many other operations on the network. For example, the right to send and receive transactions can be restricted to certain privileged addresses, so are the ability to transact and the mining privilege.

In order for a blockchain to be truly private, at least one administrator must know the real world identity of the person using the address with permission privilege. Most participants in the chain need not know each other's identities.

A key feature of blockchains is allowing peer-to-peer exchange transactions with an anonymous address so that these parties do not know the identity of its counterparty. This is as if the hotel receptionist does not disclose the room number of guests, but can deliver the message for anyone to the guests. One could imagine financial institutions transacting under many different addresses, with only regulators knowing which address belongs to which.

As a blockchain, MultiChain has the advantages over centralized database in that each participant retains full control over its assets via their private key and at the same time, the database is distributed across many entities. It is also more robust since the malfunctioning of one server will not affect the continued processing of transactions by the network as a whole.

Because MultiChain blockchain is privately owned, there is no transaction fee and block reward. However, MultiChain can be configured to use fiat currency for block rewards, minimum transaction fees, and output quantities.

Section 6.10   Quorum

Similar to HydraChain, Quorum is another private or Permissioned blockchain technology based on Ethereum

Other blockchain platforms

platform.[153] By using Ethereum's core, Quorum can incorporate most of Ethereum updates quickly and seamlessly. It uses majority voting BTF, Raft-based consensus model for faster block times, transaction finality, and on-demand block creation. It focuses on enterprise and industrial applications, which require transaction privacy and control. It is ideal for applications requiring high speed and high throughput processing of private transactions. This suits particularly well for the financial industry. J.P. Morgan is experimenting Quorum for its high speed and high throughput processing of private transactions.[154]

Much of privacy functionality resides in a separate layer on top the standard Ethereum protocol layer. Quorum treats public transactions and private transactions differently. Public transactions are visible and validated by all nodes. Only the relevant nodes can see the private transactions and execute its contracts. Other nodes, that are not part of the parties involved in the private transaction, simply skip this transaction execution. In this way, the state database splits into two databases: a private one and a public one. All nodes in the network are in tune with their public state. Nevertheless, their private state databases are selective to their involvement. However, all nodes fully replicate the distributed blockchain and all the encrypted transactions.

Microsoft's Azure Cloud Computing platform, EBaaS, hosts Quorum platform so that Quorum applications can also run on a cloud. J.P. Morgan's Quorum platform also implements zero-knowledge security layer. This gives Quorum the same privacy features of Zcash, at the same time it complies with government regulations.

Section 6.11   Hyperledger

The industry leaders in finance, banking, IoT, supply chain, manufacturing, and technology join in collaboration to develop Hyperledger blockchain frameworks. Its collaborators include well known blockchain companies (ConsenSys, R3 etc.), technology companies (Cisco, Fujitsu, Hitachi, IBM, Intel, NEC,

---

[153] https://www.jpmorgan.com/country/US/EN/Quorum
[154] https://github.com/jpmorganchase/quorum

NTT Data etc.), financial services firms, (ABN AMRO, BNY Mellon, CME Group, CLS Group, Wells Fargo Bank, etc.) and many others. The Linus Foundation hosts the collaboration. It is an open source collaborative effort created to advance cross-industry blockchain technologies.

Four Hyperledger blockchain platforms were developed for different applications: Hyperledger Fabric (a private and Permissioned blockchain platform for enterprise), Hyperledger Iroha (a version of Hyperledger Fabric focused on mobile applications), Hyperledger Burrow (an Ethereum Virtue Machine), and Hyperledger Sawtooth (Intel's blockchain platform).

Hyperledger Fabric, a private and Permissioned blockchain platform for the enterprises, is based on the earlier versions of Blockstream's libconsensus and IBM's OpenBlockchain. The production-ready Hyperledger Fabric was announced in July 2017. It has blockchain's advantages of consensus, immutability and it is modular so that it is flexible, scalable and can support plug-ins of different components to accommodate wide varieties of differences across the enterprise ecosystem. Its trust is not built on PoW, but network members are admitted through the enrollment.

To achieve the maximum flexibility, Hyperledger Fabric stores ledger data in multiple formats. It can customize consensus mechanisms and support different Membership Service Providers (MSP). [155] Members can create their own separate Mutual Distributed Ledgers (MDL) for privacy. Only members can access their MDL's.

As we have discussed in Section 6.02, like any multi-MDL platform, Hyperledger Fabric has transaction MDL and content MDL. They together form the Hyperledger Fabric's MDL. The content MDL in Hyperledger is the current state known as the "world state" and the transaction MDL contains the transaction log.

Members run applications external to the blockchain to interact with the MDL's. The apps invoke Hyperledger Fabric smart contracts written in a language called chaincode. The ledger

---

[155] "Membership Service Providers", http://hyperledger-fabric.readthedocs.io/en/release/msp.html

Other blockchain platforms

writes the consensus transactions in the order they occur. Each MDL can choose a consensus mechanism that best represents the relationships that exist between its members. The architecture of Hyperledger Fabric is highly elastic and extensible, different from other blockchain solutions. Hyperledger Fabric also supports multiple networks to manage different assets, agreements, and transactions.

Hyperledger Fabric has normal database functions such as query and update. Peer nodes validate the submitted transactions against endorsement policies described in a channel's ledger. The endorsement policies also have built-in consensus mechanisms, such as Sumeragi, a Byzantine Fault Tolerant consensus algorithm.[156] The nodes validate these transactions if they meet the policy criteria checks.[157]

Hyperledger Sawtooth Lake is another Hyperledger modular blockchain suite that supports both permissioned and permissionless deployments. It can run general-purpose smart contracts on a distributed ledger. Transaction logic in Hyperledger Sawtooth Lake is separate from the consensus layer. Consensus mechanism used is the Proof of Elapsed Time (PoET).[158][159][160][161] PoET, originally developed by Intel, is based on the fact that PoW essentially imposes a mandatory but random waiting time for leader election. Once the timer expires, the leader node can prove to all others that it has executed the "waiting step" correctly for extending the blockchain. This creates a stable consensus protocol. PoET consensus executes the waiting step in Intel's CPU, using

---

[156] yperledger Iroha's github page: https://github.com/hyperledger/iroha

[157] Hyperledger Fabric's githib page: https://github.com/hyperledger/fabric 0*T8k7vC_i_Mas85If

[158] Hyperledger Sawtooth's official website: https://01.org/sawtooth/

[159] Hyperledger Sawtooth's github page:https://github.com/hyperledger/sawtooth-core

[160] Hyperledger Sawtooth's demo on "Bringing traceability and accountability to the supply chain":https://01.org/sawtooth/seafood.html

[161] Hyperledger Sawtooth's demo on "Enabling secure and efficient bond settlement": https://01.org/sawtooth/bond.html

Intel CPU instruction set, called Intel Software Guard Extensions (SGX). This special hardware component creates an "attestation" or confirmation. Any node can use this confirmation to verify that the leader has correctly waited for the proper random time. This has the same effect as PoW does with mining.

This process is much more energy efficient than PoW. However, the probability of a node becoming the leader is proportional to the number of hardware modules under its control. As long as there is a large number of trusted nodes; PoET is compatible with permissionless blockchains.

In a Permissioned environment, the participating nodes are by invitation only, and all of them are trusted nodes. However, with known nodes, traditional PoET consensus does not offer advantages over BFT consensus protocols: BFT is more efficient, does not rely on a single vendor's hardware.

China embraces Hyperledger wholeheartedly. To date, more than 25% of Hyperledger members are from China. To enhance the collaboration with the Chinese Hyperledger community, Hyperledger Technical Steering Committee formed a Technical Working Group China (TWG China) in December 2016.[162]

The TWG China is a bridge between the global Hyperledger community, the emerging technical users, and contributors in China and the greater region, including Hong Kong and Taiwan. TWG China is now led by key contributors from China, who actively contribute to the Hyperledger Technical Community and build the consortium ecosystem. The chairpersons of the working group are from IBM, Wanda, and Huawei.

Section 6.12   Decentralized internet

Today, Internet Service Providers (ISPs) are the operators of the internet. We are paying ISP to use the internet. With the

---

[162] "Hyperledger Announces Technical Working Group China", https://www.hyperledger.org/blog/2017/01/03/hyperledger-announces-technical-working-group-china

Other blockchain platforms

blockchain technology, it is possible for us to bypass ISP and communicate through the peer-to-peer network.

The team at Open Internet Socialization Project (OISP) is working on a network called Andrena.[163] This network consists of computers in the vicinity covered by Wi-Fi network, acting as nodes in a wireless grid to help deliver data to each other. The envisioned Andrena network connects the Wi-Fi hotspots in the vicinity to form a local network. The public blockchain manages the flow of information. All users seamlessly transact with each other on the blockchain directly.

The cheap household Wi-Fi can project Wi-Fi signal only up to 1.5 miles. To reach the network beyond this Wi-Fi coverage area, it requires a higher power Wi-Fi router. There is a need for someone to install such a facility for any connectivity beyond the vicinity. Such a facility may charge for the service, but it is much cheaper than the service provided by ISP. This is illustrated in Figure 6-3.

In this way, infrastructure is decentralized. The ownership does not belong to one company, but rather the users. This is another example of the sharing economy. Instead of paying ISPs for providing the infrastructure, the community pays each other, which is usually much cheaper. The blockchain technology ensures fairness without trust. It allows Andrena networks to form and organize, without a central authority.

Of course, some nodes in the network will be gateways to the Internet outside of the Andrena network. These are users that maintain connections to the existing ISP providers. The price of the Internet at this level is much cheaper than what we pay for today via ISPs.

There are many technical issues to be resolved, such as to verify that nodes are providing their agreed amount of bandwidth. Security is another issue. An encrypted tunnel between users' devices and the gateway nodes needs to be created, ensuring that no one in the middle can see the data traffic.

---

[163] "Decentralized internet on blockchain", https://hackernoon.com/decentralized-internet-on-blockchain-6b78684358a

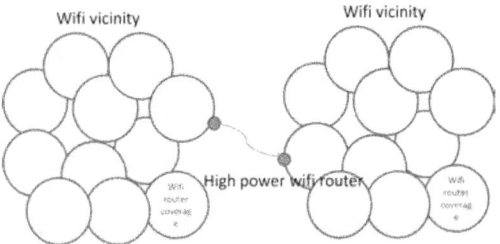

Figure 6-3 A P2P internet

Section 6.13   Other blockchain platforms

There are many other blockchain platforms. Their numbers are increasing every day. Openchain [164] is an open source distributed ledger system for issuing and managing digital assets. Tokens on Openchain can be pegged to Bitcoin, making it a sidechain. The consensus mechanism used by Openchain is the Portioned Consensus. Openchain is a very centralized blockchain. Different authorities validate different transactions depending on the assets being exchanged. However, each transaction is validated by only one authority. Like Hyperledger, each owner controls his or her own Openchain MDL. However, different MDLs can connect to each other. This is more like the traditional banking system.

Stellar is another open-source, distributed payment infrastructure that connects banks and payment systems. Stellar enables the building of mobile wallets, banking tools, and smart devices. Application Programming Interface called Horizon is the user interface, which connects users to the Stellar Core, the backbone of the Stellar network. The consensus mechanism is the Stellar Consensus Protocol.[165]

BitSE's Qtum is a new operating system for blockchain development combined with a modularized basic chain, which

---

[164] https://www.openchain.org/
[165] "Stellar Consensus Protocol": https://www.stellar.org

Other blockchain platforms

combines the advantages of numerous public chains like Bitcoin and Ethereum.

DGX[166] is a new gold-backed Ethereum smart contract coin developed by DigiX. Each GDX is equivalent to 1 gram of gold. One can own, save and transact Gold in GDX.

Decentralized Capital, a UK company, developed DC Assets[167], which can exchange to Dollars, Euros, and six others fiat currencies. DC Assets is a digital currency secured by the Ethereum network and collateralized by customer deposits. Users can transact with DC Assets for products and services on the Ethereum blockchain.

It seems that the potential applications are limited only by imagination. For example, Microsoft is working on a secure ID system to prevent human trafficking. Everledger[168] is developing a system to track the origin of diamonds. Imogen Heap just released her song using a blockchain platform, which allows artists to sell their music directly to their fans. Japan and Sweden are experimenting to put their real estate registration on blockchain to improve efficiency and to prevent fraud. And the list goes on and on.

Goldman Sachs[169] thinks that blockchain could disrupt everything. The blockchain technology, also known as the Mutual Distributed Ledger (MDL) technology, will have immense potential to transform the way people and business to handle identity, transaction, and asset, as we shall see in many disruptive applications discussed in this chapter.

The most important characteristics of the MDL is that it can replace the trust, which is a necessary element to transact business in today's world. It can do so with three core functions that trusted third parties perform: validation, transaction, and

---

[166] https://digix.global/
[167] https://www.decentralizedcapital.com/#!/
[168] https://www.everledger.io/
[169] "What if I told you", Goldman Sachs, http://www.goldmansachs.com/our-thinking/pages/macroeconomic-insights-folder/what-if-i-told-you/report.pdf

recording. Everyone shares the ledger. The computer systems follow a common protocol to add new transactions.

In addition, MDL is available globally and instantly, verifiable and secure, and can do so cheaply and robustly. This is why the blockchain technology opens the door to infinite possibilities of business transactions.

Historically, distributed ledgers have suffered from two problems: insecurity and complexity. However, with the arrival of blockchain technology, these two problems are eliminated by employing cryptography, and its integrity to prevent double spending is maintained by block validation mechanism. The simplicity and robustness of the blockchains ensure that the system is simple to understand and to deploy.

MDLs incorporating trusted third parties had significant potential in financial services, such as know-your-customer (KYC), antimony-laundering (AML), insurance, credit and wholesale financial services. Blockchain technology has proven to be robust. It has gained the confidence from such large companies such as Nasdaq, BNY Mellon, UBS, USAA, IBM, Samsung and many others to turn to blockchain for their future applications. MDLs can provide many competitive advantages. In this chapter, we are going to discuss some applications and potential applications in development, besides the digital currencies that we have already discussed in detail.

Section 6.14 Beyond blockchain

The blockchain is not the only technology that can construct an MDL. Hashgraph is touted as an alternative to the blockchain to build MDL technology. Hashgraph can resolve Bitcoin's scaling and security issues, while also pushing the use of distributed consensus applications into new areas. According to Hashgraph, the company which pioneers the technology, Hashgraph is a data structure and consensus algorithm that is faster, fairer, and more secure than blockchain. [170] It uses two

---

[170] https://hashgraph.com/

Other blockchain platforms

special techniques - Gossip about Gossip and Virtual Voting.[171] Gossip about Gossip involves attaching a small additional amount of information to a pair of hashes (Gossip) that contain the last two people talked to. By doing this, a Hashgraph can be built and updated whenever additional information is gossiped, on each node.

Hashgraph deploys asynchronous Byzantine process to achieve agreement in the MDL.[172] It makes no assumptions about how fast messages are passed over the internet. This makes it resilient against DDoS attacks, botnets, and firewalls. The Hashgraph algorithm works without the PoW, and therefore, does not require extensive resources. It can deliver low-cost and high-performance processing. For example, Bitcoin can process at 4.17 transactions per second on average. While Hashgraph transaction rate can go as fast as 250,000 transactions per second.

In addition, Hashgraph allows a fairer system of operations. In the blockchain system, miners can pick and choose any unconfirmed transactions to be included in a block, usually the high fee-paying- transactions, and can delay low-fee-paying transactions indefinitely. Hashgraph utilizes consensus time-stamping and prevents any node from changing the consensus order of transactions by denying the ability to manipulate the order of transactions.

Despite its obvious benefits, as a latecomer, Hashgraph has a long way to go before it can boast the popularity enjoyed by blockchain technology. However, in the ever-evolving world of MDL technology, it seems that the next paradigm shift is never far away.

---

[171] "Hashgraph vs. Blockchain Is the end of Bitcoin and Ethereum near?",George Kingslay, https://coincodex.com/article/1151/hashgraph-vs-blockchain-is-the-end-of-bitcoin-and-ethereum-near/

[172] "Asynchronous Byzantine Agreement Protocols", Gabriel Bracha, https://www.sciencedirect.com/science/article/pii/089054018790054X

Chapter 7 Financial blockchain applications

Section 7.01 Banking and payment

Besides digital currency, the banking and payment are the most obvious applications of blockchain technology. In fact, a completely new branch of technology called financial technology, or Fintech for short, is springing to life. It envisions that Fintech will be able to provide financial services to billions of people around the world, circumventing the current financial establishment. Blockchain can transform the financial services industry by making transactions faster, cheaper, more secure and transparent.

By the end of 2017, 15% of the large banks employed one kind of blockchain technology or another. Their goal is to leverage blockchain technology or MDL's to create a decentralized system to replace trust to significantly slash all types of transaction fees and reduces processing times.

However, the term "decentralized" is deceptive, as we have seen in the discussion of national digital currency. Although blockchain attracts the attention as a decentralized system, by construction, it can also be a very centralized system or anywhere in between. In fact, blockchain applications can be decentralized in its deployment, but still, maintain centralized authorities.

Banks are experimenting with the distributed ledger approach to create efficiencies without giving up their control of the financial establishment. Their goal is to automate processes,

Other blockchain platforms

speed up the transaction, reduce the cost, and enhance data security. To deploy blockchains, banks need to review the existing complex financial structures to determine which processes to move to a blockchain. The long-term plan of action will not be easy because blockchain technology itself is evolving. Furthermore, since the blockchain technology is so new, a transition from centralized legacy to fully distributed digital transaction processing may encounter unexpected glitches. The subject is large and complex.

In November 2015, the Council of the European Union approved the European Payment Service Directive version 2 (PSD2) legislation created by the European Parliament in response to the upcoming digitalization of the financial services. It provides a basic legislative framework for online payment, the development and use of innovative online and mobile payments and cross-border European payment services. It is a step towards a digital single market. PSD2 became effective on 13 January 2018. Britain responded with similar plans called Competition and Markets Authority (CMA). With the legislative in place, many European companies are rushing to develop the online payment service platforms.

For example, Moneybookers/Skrill, a London based company specializes in the transfer of foreign exchange into domestic banks in e-banking. Geoswift[173] is involved in cross-border payment platform. ChinaPnR[174] provides payment services including online payment/fund management/POS payment/mobile payment and tailored integrated payment solutions. All in Pay[175] engages in off-line payment.

Elsewhere in the world, there are many activities for the same market space. For example, Foxconn, the world leading contract manufacturer, invested in a Californian blockchain start-up called Abra together with other investors include Arbor Ventures, American Express Ventures, Jungle Ventures, Lehrer Hippeau and RRE. Abra raised a total of US$35 million. Abra is in the mobile P2P consumer payment service using Bitcoin. Its

---

[173] http://www.geoswift.com/our-company/
[174] http://en.chinapnr.com/
[175] http://www.allinpay.com/

app using blockchain technology can transfer cash between two smartphones. The UK has seen its first online-only bank – the Atom Bank in 2016. Within one year of its opening, it has broken the record of 5,000 new customers' signups in one day. [176]

India also developed its own mobile-first financial services company, Paytm. [177] It offers services for payments, banking, lending and insurance to consumers and merchants. Paytm uses QR Code for the transaction. Paytm can count on 7 million merchants across India to accept payments directly into their bank account without a transaction fee. In 2017, the company launched the Paytm Payments Bank to bring banking and financial services access to half-a-billion un-served and under-served Indians. In 2018, it set up Paytm Money to offer investment and Wealth Management products.

The cross-border payment is another market growing with the tourist traffic across the globe and cross-border retail business. PayPal is catering to Chinese consumers by collaborating with UnionPay to support outbound cross-border transactions and payments. This enables Chinese consumers to purchase goods from foreign e-retailers directly in countries where UnionPay still has limited acceptance. Merchants in European countries can also accept payments from UnionPay credit and debit cards through PayPal's Braintree platform for mobile commerce. PayPal and UnionPay also launched "PayPal China Connect", allowing Chinese customers to convert RMB to their payment currency of choice directly.

Section 7.02   Credit card and loan applications

The credit card companies do not want to be left behind by the blockchain technology. Blockchain technology may seem to be a threat to these companies that it provides P2P transactions bypassing intermediates; however, they can also be big beneficiaries to use the new technology to streamline their operations. The payments industry as a whole is undergoing a

---

[176] https://en.wikipedia.org/wiki/Atom_Bank
[177] https://paytm.com/

Other blockchain platforms

paradigm shift. Moreover, the industry is big: the global consumer payment market reached US$47 trillion in 2014, the first year that the share of digital payment volumes exceeded that of paper-based payment methods.

Financial institutions are keen to attain Straight-Through-Processing (STP) concept.[178] STP is a scheme to optimize the speed of transactions. It allows the transferring of electronically entered data from one party to another in the settlement process without manually re-entering the same data repeatedly over the entire sequence of events. One of the benefits of STP is a decrease in settlement risk and time.

To do so, it is necessary to streamline the process of transactions across multiple points. By allowing information to pass along electronically, it requires the manual entry of data only once at the source. Multiple parties can receive the same information simultaneously if needed. STP will eliminate human intervention from financial transactions, thus, reducing the cost of transaction as well. A true STP will save tremendous cost in the credit card processing.

Blockchain technology, ideal for digital transactions, has the potential to implement STP and to transform the global financial network. The technology could accelerate the velocity of money, and provide a path for legacy banking systems to interoperate, greatly improving efficiency. The use of distributed ledgers has the potential to disrupt the payment industry in the near future.

Along with VISA, other credit card companies, like MasterCard, American Express and banks and technology firms are all feeling the heat to adopt blockchain technology due to a potential reduction in costs, and improved product offerings. If they do not swim, they will sink.

VISA provides financial institutions with payment and money transfer products such as credit card, debit card, prepaid card, commercial and mobile money transfer. The company

---

[178] "Straight Through Processing", https://www.investopedia.com/terms/s/straightthroughprocessing.asp#ixzz51TkRyllZ

operates a proprietary transaction-processing network, called VISANet. VISANet facilitates authorization, clearing, and settlement of payment transactions worldwide while offering fraud protection for account holders and rapid payments for merchants. It is capable of handling 24,000 transaction messages per second. It is a leading choice for cardholders and financial institutions worldwide.

VISA is kicking off its blockchain development. Recently, it issued the popular Bitcoin debit card, the Coinbase Shift Card. It considers Fintech as a unique opportunity at a time when the payments industry is undergoing a digital transformation. To start with, VISA Europe Collab is working with Epiphyte, a European startup, to utilize the blockchain to facilitate international money transfers. The project created a prototype smartphone wallet app and to enable the processing of remittances directly from a VISA card to a destination payment source over the Bitcoin blockchain. VISA has also invested in a blockchain developer platform, Chain.com, which serves an enterprise market. The incorporation of blockchain technology into the VISANet operation can potentially be very beneficial to VISA. Other credit card companies are not far behind.

Kreditech is a Germany based Fintech company.[179] It uses non-traditional data sources and machine learning to provide financing to the people with little or no credit history. It is based on the conviction that not all the people without credit history carry high financial risk. By the use of its proprietary credit decision technology, Kreditech is doing good business to help people in financial needs and taps into a largely untapped market of the financially underserved population.

In summary, the blockchain-based networks and traditional technology-based alternative payment solutions will continue to merge. The payment service market landscape will be dramatically different a decade from now.

Section 7.03   Insurance

---

[179] https://www.kreditech.com/

Other blockchain platforms

Insurance is a big business. The worldwide insurance market in 2015 is around US$4.5 trillion.[180] Today, the insurance industry is built on the trust. For example, people who buy long-term care insurance must trust the insurance company to pay the cost of long-term care when needed. An insurance policy is really a contract between the insurance company and the insured.

The global insurers are quick to embrace the blockchain technology. The insurance companies can benefit from the blockchain technology in many aspects. First, blockchain can simply the insurance administrative and process functions. However, the most important application of blockchain for insurance is coming from the nature of the insurance contracts, in which certain specific events trigger the obligations. The insurance policy is essentially a contract to transfer risk from the client to the insurer.

The insurance industry has a set of complex legal and regulatory framework, addressing issues of ownership, responsibility and potentially jurisdiction and dispute resolution. Therefore, the insurance contracts are usually complex, to develop smart contracts for the insurance.

The insurance industry has launched the Blockchain Insurance Industry Initiative or B3i project[181] in October 2016. B3i is dedicated to developing trading platforms across the insurance value chain using blockchain technologies. It aims at efficiency improvements in transacting insurance. Using blockchain to handle insurance contracts eliminates the need for multiple databases and the errors that arise from maintaining and transferring data among them.

B3i has 38 members of insurers, brokers, reinsurers. Its membership includes big insurance companies, such as Allianz, Swiss Re, Liberty Mutual, Sompo Japan Nipponkoa, Reinsurance Group of America, Hannover, Generali Group, SCOR, and others. The market share of B3i members is around 43% of ceded and

---

[180] "Global insurance market trend", EY report, http://www.ey.com/Publication/vwLUAssets/ey-global-insurance-trends-analysis-2016/$File/ey-global-insurance-trends-analysis-2016.pdf

[181] https://b3i.tech/home.html

70% of reinsurers' premiums worldwide. The target date to release the platform for commercial use is by 2018.

Besides B3i, there are also other consortia in the same space: RiskBlock[182], EY /Microsoft /Maersk, and R3.[183] The EY /Microsoft /Maersk project aims for marine insurance but can be applicable to another insurance market as well in the future. The blockchain platform will reside on Microsoft's Azure. The distributed ledger will be used to capture information about shipments, risk, and liability, and to help firms comply with insurance regulations. It will also ensure transparency across an interconnected network of clients, brokers, insurers and other third parties.

Symbiont developed an MDL platform to create smart insurance contracts that can execute payments with little or no human involvement. MDL based smart insurance contracts can automatically authorize and execute the required payment when the condition meets the contractual parameter. MDL is also safer than current database technology because of the encrypted MDL data. MDL platform also allows an individual to establish a global identity protected by encryption but immediately available to individuals and organizations authorized to access the information.

Section 7.04   Security trading

The potential for added efficiency in share settlement makes a strong use case for blockchains in the security trading. When executed peer-to-peer, trade confirmations become almost instantaneous (as opposed to taking three days for clearance). Potentially, this means intermediaries — such as the clearinghouse, auditors, and custodians — get removed from the process.

---

[182] https://www.theinstitutes.org/guide/riskblock

[183] "R3's partnership with ChainThat is one gaint leap for insurance", https://www.the-digital-insurer.com/blog/insurtech-r3-chainthat-partnership-giant-leap-insurance/

Other blockchain platforms

Numerous stock and commodities exchanges are prototyping blockchain applications for the services they offer, including the ASX (Australian Securities Exchange), the Deutsche Börse (Frankfurt's stock exchange) and the JPX (Japan Exchange Group). Most high profile because the acknowledged first mover in the area, is the Nasdaq's Linq, a platform for private market trading (typically between pre-IPO startups and investors). A partnership with the blockchain tech company Chain, Linq announced the completion of it its first share trade in 2015.

More recently, NASDAQ announced the development of a blockchain project for proxy voting on the Estonian Stock Market. The Australian Securities Exchange (ASX) has been testing blockchain in equity trade settlement and clearing. Nonetheless, Australia's equity market is dematerialized and relatively small making blockchain adoption potentially more straightforward.

A number of financial institutions are participating in working groups as a means to develop blockchain further. An example of such a group is the R3 Consortium, which comprises a group of banks looking to develop blockchain technology across financial services. Reports also indicate several high-profile asset managers are exploring blockchain as a mechanism to speed up transactions in illiquid assets.

The same STP concept for credit card process applies to security trading as well. Blockchain has the potential to disrupt capital markets. Potentially, it could save between US$15 billion and US$20 billion by 2022 through streamlining cross-border payments, securities trading, and regulatory compliance.

Some challenges must be overcome before this can be realized. It will require harmonized standards and regulation agreed by the industry, regulators, and governments. The scale of this challenge is huge. However, once overcoming the challenges, the blockchain technology has the potential to impact markets globally including emerging economies, which are in the early stages of developing their market infrastructures.

For the security trading application, the system must be capable of recording transactions in an immutable ledger where transactions are signed cryptographically and impossible to forge. Blockchain technology can meet this requirement. Domus Tower

Blockchain,[184] developed by a startup in San Francisco, intends for applications such as securities trading. [185]

Domus Tower Blockchain can process over 1 million transactions per second on Amazon's Web Services with the potential to scale to greater than 10 million transactions per second. Data storage is contained in a Merkle Directional Acyclic Graph (Merkle DAG), a variation of DAG whose data structure is similar to Merkle tree. The data transmitted to the blockchain is digitally signed and verified before it is written to a block. It has the advantage over the traditional security trading platforms that it is faster and more secure.

As a blockchain technology platform, Domus Tower provides an immutable, permanent record of time-stamped transactions. All transactions are audited cryptographically in real time with a Merkle root. Every transaction is signed in real-time with a public/private key signature that is impossible to forge. The append-only, time-stamped structure of a blockchain makes it much easier to scale. These qualities allow Domus Tower platform to excel over the traditional database security trading systems, while still work in the same frame of the securities industry today where parties know each other and trust data feeds. It also excels over the Permissionless blockchain platforms, which has scaling and transaction speed problems.

Domus Tower Blockchain can create linked blockchains where the assets of an account on one blockchain must match the liabilities on the account of another blockchain. It is a centralized, Permissioned blockchain. All transactions in the Domus Blockchain require digital signature. The cryptographic digital signature is verified before the data is written to a block. The "Merkle root" guarantees an immutable history of all signed data stored in all the prior blocks of the entire Merkle graph. Since the Domus Blockchain is a Permissioned blockchain, it does not use PoW/PoS or Byzantine Fault Tolerance. It operates in an environment where participants trust each other. Only the agents

---

[184] http://domustower.com/
[185] "Domus Tower Blcokchain", Rhett Creightoon, http://www.domustower.com/domus-tower-blockchain-mar-22.pdf

Other blockchain platforms that have access to a blockchain can write transactions to that chain.

Domus Blockchain performs a series of micro-services, which does a specific task. Micro-services operate on simple data-driven interfaces. The major components include a "signature verifier", "transaction batcher", "block maker", and a "client". The Domus Blockchain achieves high scalability. It does not use a vote-based approach to writing transaction data.

Section 7.05 Commodity trading

Today, the future exchanges provide a market for trading commodities using future contracts. Buyers promise to buy tons of coffee, for example, by signing a future contract with the supplier for the future delivery of a certain lock-in price. However, from the signing of the contract to the actual delivery, many things can change.

Trading commodities using blockchain smart contracts can have all the benefits of the blockchain advantages. Mercuria, the Swiss-based commodities giant with annual turnover $91 billion, successfully tested an oil trade using blockchain technology. [186]

Potentially, using blockchain to trade commodities can save billions of dollars a year in commodities trading by scrapping millions of paper documents and moving to a digital equivalent with blockchain. However, the implementation may require a drastic change in the way that conducts the commodity trading.

There is more than one approach to trade commodity using blockchain technology. The most direct approach is to use the smart contracts, which still settle the trades in the fiat currencies. A more radical approach is to digitize the commodity assets. For example, a certain number of tokens can be the digital equivalent of one ton of coffee of a certain quality. These tokens represent the

---

186 "Mercuria introduces blockchain to oil trade with ING, SocGen", Dmitry Zhdannikov, Reuters, https://www.reuters.com/article/us-davos-meeting-mercuria/mercuria-introduces-blockchain-to-oil-trade-with-ing-socgen-idUSKBN1531DJ

ownership of one ton of coffee of this quality, and they can redeem the coffee. They can trade for other cryptocurrencies or fiat currencies.

ACChain has digitized Puer tea and has plans to digitize world's grain supply.[187] For such digitization to happen, IoT must be a mature technology. We will illustrate by an example of soybean transaction: Brazil ships 100 tons of soybean to China. When the shipment arrives at the port of destination, the IoT with sensors will first verify the tonnage, moisture content, the color of the soybean to ensure all the physical parameters are within the specifications of the smart contract. If all the conditions are meeting the contract terms, the smart contract will trigger the payment event. It is not up to the buyer to accept or not accept the merchandise. Under certain unsatisfied conditions specified in the contract, the smart contract will automatically apply a discount to the price of the soybean. In the worst case, the smart contract will reject the shipment, and will not trigger the payment. Likewise, smart contracts for any commodity – ore, petroleum, wheat, rice, fruit, etc, can apply similar criteria.

It is easy to establish the specifications for degradable products, such as fruits, fresh produce, for the shipment, so that if there is a refrigeration problem during the shipment, the smart contract of shipping will apply a certain penalty to the shipper.

When the smart contracts using blockchain for commodity trading becomes prevalent, they can be easily standardized. These predefined smart contracts are executed by all the relevant counterparties with conditions for execution clearly spelled out.

During the contract execution, the events trigger actions, such as the transactions initiated, information delivered, etc. For the digital assets on the chain, the smart contract can settle the account atomically, in other words, agreed by all parties. To settle the assets off the chain, the smart contract needs to include off-chain settlement instructions. Smart contracts guarantee a very specific set of outcomes. Energy and commodity trading on blockchain will drastically change the way trading is performed

---

[187] https://www.acchain.org/en/

Other blockchain platforms

and will result in a great saving in the transactions in billions of dollars.[188]

Section 7.06   Energy trading

Energy is a special form of commodity, so as the energy trading, especially the trading of electricity. With the wider use of renewable and clean energy, such as rooftop solar energy installations, micro-generation of electricity is becoming a huge trend in the power generation business. Households are not only consumers but also they become small energy suppliers. However, household energy supply is erratic depending on the weather and consumption amount. It is very difficult to balance the supply and demand in each unit of the micro-generation. In such a blockchain-based platform, energy is treated as an asset. This is why blockchain technology is ideal to manage such a situation.

Currently, the surplus energy generated by households is sold back to the grid with much lower price than the electricity from the grid. The application of blockchain technology to the micro-generation of energy becomes a great equalizer. Smart meters register electricity produced and consumed in a blockchain. It then allows for consumption of the surplus energy in the neighboring household, which pays to the original producing household. The credits can be redeemed against the grid for his future energy use. In this way, the surplus energy is consumed near the household that produces energy without the need to travel a great distance.

The blockchain enforces these contracts in real time automatically and without supervision, allowing creating a utility market with minimal effort. It is foreseeable that community can install small-scale public power generation utility in the neighborhood and become self-sufficient in the energy consumption. It can cut down the transmission loss dramatically.

---

[188] "Overview of blockchain for energy and commodity trading", Ernst & Young LLP,
http://www.ey.com/Publication/vwLUAssets/ey-overview-of-blockchain-for-energy-and-commodity-trading/$FILE/ey-overview-of-blockchain-for-energy-and-commodity-trading.pdf

The large-scale application of the blockchain smart contracts in the micro-generation of energy can not only provide the households with better energy selling price but also will avoid long distance energy transmission, smooth out the energy utilization across the grid and encourage the installation of localized micro-generations. In a sense, for the first time since the invention of the large-scale electric power plant, the electricity generation can be decentralized.

The blockchain allows the purchase and sale of electricity according to user's personal preferences and needs. By doing so, it added monetary benefits through optimizing for the most favorable energy transactions at any given time and it Increases independence from the grid in case of power supply issues. In order to develop and enable P2P trading of electricity, micro-generated electricity needs to be verified, with the generation time and the amount recorded. This is to monetize the energy as the price of electricity varies over time based on supply and demand. The monetization is done by using tokens, which can be traded and audited in a blockchain.

The word P2P trading of the electricity may be misleading. Because of the small amounts of electricity surplus by domestic suppliers is intermittency, instead of sourcing electricity from a single household, consumers will have to buy and sell electricity across an open market, swapping their energy supplier on a minute-by-minute basis. Households can buy electricity based on their own personal preference, such as clean energy, nuclear or best price. Essentially, it is a bazaar of energy market on the grid.

In the US, the software foundry company, ConsenSys, is developing a range of applications for the electricity market. One of the projects called Transactive Grid[189][190] is in partnership with

---

[189] "P2P energy transaction and control", https://www.slideshare.net/JohnLilic/transactive-grid

[190] "Enerchain: A Decentralized Market on the Blockchain for Energy Wholesalers", Morgen Peck, https://spectrum.ieee.org/energywise/energy/the-smarter-grid/enerchain-a-decentralized-market-on-the-blockchain-for-energy-wholesalers

Other blockchain platforms

a distributed energy outfit, LO3.[191] The project uses Ethereum smart contracts to automate the monitoring and redistribution of micro-grid energy. Ethereum-based smart contracts automatically redistribute the energy.

Likewise in Europe, twenty-six European energy trading firms jointly conduct P2P trading in the wholesale energy market on a platform called Enerchain. A German blockchain technology company Ponton developed the platform.[192] This platform enables traders to buy and sell directly without broker and exchange. However, the blockchain based energy trading is not limited to P2P. It can be B2B as well. Ponton's Enerchain platform also works for the B2B.

The large power grid operators are also looking into the potential for blockchain technology to simplify the power grid management processes. The complex task of electricity distribution is managed by many specialized organizations. A Transmission System Operator performs the balancing of energy to keep grid load and frequency stable. A Distribution System Operator monitors the loading on the grid and keeps it stable at the local level. They need to coordinate any supply and demand in-balance and disruption, such as outages, congestion as well as micro-generation in the local grid. There are also aggregators, who pool many power plants to balance the power in the grid to make sure that the flow is continuous and constant. Ponton has also developed a blockchain technology-based platform for real-time grid management, called Gridchain, which is ready for the field-testing.[193]

Section 7.07 The sharing economy

---

[191] https://lo3energy.com/
[192] "Enerchain P2P trading project",
https://enerchain.ponton.de/index.php/21-enerchain-p2p-trading-project
193 "Gridchain – blockchain based process integration for the smart grids of the future",
https://enerchain.ponton.de/index.php/16-gridchain-blockchain-based-process-integration-for-the-smart-grids-of-the-future

Companies like Uber and AirBnB are flourishing. Evidently, the sharing economy is a successful business model. Currently, however, users who want to hail a ride-sharing service have to rely on an intermediary like Uber. With the arrival of blockchain technology, by enabling peer-to-peer payments, the blockchain opens the door to direct interaction between parties — a truly decentralized sharing economy results.

OpenBazaar uses the blockchain to create a peer-to-peer eBay. Download the app onto your computing device, and you can transact with OpenBazaar vendors without paying transaction fees.

Before the automobiles became popular in China, bikes were the major transportation tools, a ubiquitous image in the Chinese cities. Now, the bikes are making a coming back in a different way. People ride shared bikes. The bike sharing business is booming. Several Chinese startups have deployed tens of millions of colorful bikes in cities across the country. The rental costs as little as 15 cents for 30 minutes. Users can pick up and drop off bikes everywhere. It is convenient, cheap and ecologically friendly, and all you need to rent a bike is a smartphone app. Ofo and Mobike are the most successful of the bike-sharing companies to date and are now valued at more than $1 billion.

Peer-to-peer markets, collectively known as the sharing economy, have emerged as alternative suppliers of goods and services traditionally provided by long-established industries. Not only bikes are shares, even houses can also be shared. The home sharing business AirBnB[194], which has 300,000 shared homes in 191 countries, makes its impact felt in the hotel business.

Section 7.08   Peer-to-peer lending

ROSCA stands for Rotating Savings and Credit Association. It is a popular type of P2P lending. Billions of people over the world practice the traditional form of P2P lending since

---

[194] https://en.wikipedia.org/wiki/Airbnb

Other blockchain platforms

ancient time.[195] A participant can borrow from the pool of money collectively saved for anything, buying a car, down payment for a house, etc.

A ROSCA is organized by a group of trusted parties, who agree to contribute a fixed amount of money at a fixed interval, called "round". At the end of each round, the pot contains the sum of everyone's contribution. Users bid on this amount in a reverse auction manner, where the lowest bid wins. The winning bidder then receives his bid. The remaining will be split equally among all of the participants as a form of interest payment. If a bidder's desire to get the money is strong, he will have to enter lower bid to ensure that he would win. In effect, he is paying higher interest.

In each round, there is only one winner. Each user is guaranteed to win one round in turn. A winner will have next opportunity to bid only after all the parties receive their winning bid once. This ensures a Pareto optimal outcome, where everyone is better off as they would have been had they saved the money alone.

ROSCAs are seeing growing popularity in the US, primarily within immigrant populations. It is also becoming popular in places where informal credit is the norm, such as China. Community systems worldwide are another reason why blockchain is great for leveraging reciprocal finance.

The traditional ROSCA requires trust. The scope of ROSCA is limited to a small group of people who know each other and trust each other to be participants. Therefore, its scope is limited. With the arising of blockchain technology, suddenly people find that ROSCA scheme can be extended far beyond the familiar circle because trust is no longer an issue. Blockchain provides a perfect platform for ROSCA application.

ROSCAs can be a successful supplement to regular financial services — another tool in the toolbox. Reciprocal aid is a perfect fit for blockchain technology, enabling people around the world access to more advanced financial tools. Most of all,

---

[195] " Indigenous savings & credit societies in the developing world" F.J.A. Bouman, ,Rural Financial Markets in the Developing World World Bank, Washington, 1983

ROSCA is a self-financing model. All the profits are distributed to the participants. The idea for ROSCA is to have a community of like-minded people with financial goals.

Anyone can start a ROSCA with ROSCA blockchain application, and sets rules: number of maximum participants, the time period of the round, the amount of contribution. It can also be dissolved when all the participants receive the bid.

All kinds of platforms with ROSCA concept, like eMoneyPool, Monk, Puddle, Moneyfellows, ROSCA Finance, Partnerhand, StepLadder, WeTrust, etc are springing up. One of the most successful platform WeTrust raised $6.5M market capital in an ICO in April 2017. Its market capitalization has tripled within 10 days of coin issuance. As a peer-to-peer financing system, the platform has to be Permissionless.

Section 7.09   Cross-border retail business

Cross-border trade used to be the arena of business-to-business. The amount of cross-border retail purchase is negligible. This is because of the hassle of shipping, custom, currency exchange, most of all, the guarantee of receiving money. Now it is changing. With the e-commerce boom, retail customers can order products directly from overseas. Cross-border e-commerce is booming, and there is no end in sight. Cross-border e-commerce trades will more than double over the next five years reaching US$424 billion by 2021. It will make up 15% of all online purchases in 2021.

China is the driving force behind such tidal change. Its share of the online cross-border market will grow to 40% by 2021. With the ease of placing orders online for overseas products, Chinese consumers are ever more attracted to buy foreign gadgets not available in China. Because cross-border shopping not only offers customers better prices, it can also provide a degree of protection from fake or counterfeit goods by buying directly from the well-known merchants.

When Jack Ma, founder of Alibaba, met with Trump in 2017, he promised to use Alibaba online platform to sell American

Other blockchain platforms

goods to the Chinese consumers. In the "Small Business Summit" held in Detroit by Alibaba in June 2017 attended by thousands of small business owners across America, Jack Ma laid out a plan for American small businesses to reach Chinese consumers directly.

Major U.S. brands and retailers have been successfully selling to Chinese consumers through Alibaba's Tmall and cross-border e-commerce platform Tmall Global for years. However, this is the first time such a platform is available to small businesses. With the infrastructure like e-commerce platform, there are no more barriers to the distance or even borders for doing retail business.

However, today, besides the free trade agreement between nations, the cross-border retail business is still full of regulatory and logistics issues. Alibaba is discussing with the World Trade Organization to roll out a global trade platform – eWTP – which could expedite, simplify and substantially increase the volume of cross-border e-commerce transactions. It could help millions of small business that previously missed the benefits of free trade agreements, largely due to the excessive costs of compliance and the confusing assortment of rules, regulations and red tape.

When small businesses have direct access to consumers, they can bypass middleman (exporters, importers, whole sellers etc.). By doing so, they can reap the benefit of cost savings. Jack Ma proposed eWTP, the e-World Trade Platform, in the 2016 G20 meeting in Hangzhou. It is the first international e-commerce platform aimed at Small and Medium-sized Enterprises or SME. For the first time, SMEs have the same facility to reach worldwide consumers like the giant companies using this platform. Its objective is to enable small and medium-sized businesses to reach consumers worldwide without going through the intermediates.

Alibaba launched the first eWTP known as DFTZ (Digital Free Trade Zone) in Malaysia in February 2017.[196] It is an online free trade zone, employing technology to reduce the barrier to trade. It enables Malaysian SMEs to expand their business and export globally and positions Malaysia as a regional hub for e-Commerce logistics and the preferred gateway of choice for global brands and marketplaces into ASEAN. In another example,

---

[196] https://mydftz.com/dftz-goes-live/

Alibaba collaborated with Philippine telecommunication company Globe Telecom to offer real-time cross-border remittance services using blockchain technology for individuals and small and medium-sized enterprises. The new service allows users to send and receive money within seconds across boarder.

To spread eWTP worldwide requires agreement among the countries. The technology is already available today. The domestic e-commerce technology can be easily adapted for the international DFTZ. However, due to the foreign exchange, customs duty and cross-border transactions, additional capabilities are required on the platform. Blockchain technology is ideal for such application. Ant Financial, an Alibaba subsidiary, is currently exploring blockchain technology to simplify and streamline its financial processes.

By reducing national barriers to trade, and applying new technologies, Ma hopes that any small business can utilize a smart-road of smart-hubs to sell their goods and services around the world. Ant Financial bade to acquire MoneyGram in 2017 for US$880 million to be the worldwide transaction settlement platform for eWTP. Unfortunately, the Committee on Foreign Investment in the United States ("CFIUS") blocked the merger in early 2018. Nevertheless, even without the merger, MoneyGram and Ant Financial formed a new strategic business cooperation to explore and develop initiatives to bring together their capabilities in remittance and digital payments to provide their respective customers with user-friendly, rapid-response and low-cost money transfer services. This move would give Alibaba an immediate access to the establishment, which handles cross-border currency transfers in nearly 200 countries.

Section 7.10   Counterparty platform DApps

We have discussed in Section 5.11 that Counterparty platform can run smart contracts on Bitcoin blockchain. Real world assets can back up the digital assets on Counterparty platform. Some interesting applications are:

- Betting - Counterparty turns the Bitcoin blockchain into a betting platform, which can predict the market. Users can place bets on the information broadcasted. The

Other blockchain platforms

Blockchain protocol escrows funds automatically and stores it securely in the Bitcoin blockchain. Funds placed on bets are inaccessible until the bet is resolved or expires. Broadcasters can set a fee fraction to receive for their betting feeds as an incentive to run their broadcasts.

- Token Controlled Access (TCA): Token Controlled Access is to have access to private events, such as music events, parking tickets etc. based on the ownership of tokens. In real life, tickets are tokens. People buy tickets to cinemas, flights, etc. Counterparty tokens are publicly tradable and have a monetary value.
- Proof of Publication: Using broadcasts, users can publish timestamped information onto the Bitcoin blockchain and such timestamp cannot be deleted or altered. This makes it possible to verify publisher's claim of the time of publication of a piece of information such as patent. It will find wide use to protect patents, trademarks and any other time sensitive information.
- Artist work coin: In the music world, the use of blockchain technology makes it possible to use music coins, such as the Tatiana Coin to buy songs directly from the singer. Singer-songwriter Tatiana Moroz released one of the earliest digital artist tokens. People can use Tatiana Coin to redeem for her products and services.[197]
- Crowdfunding: Crowdfunding platforms powered by blockchain technology remove the need for the trusted third party. They allow startups to raise funds by ICO. One can issue a Counterparty asset in a crowdfunding event to raise funds for a project. Doing so will inspire more trust than the regular crowdfunding because it is bounded by a smart contract. Swarm, Konify, and Lighthouse are three decentralized crowdfunding platforms.[198]
- Voting: Counterparty supports voting using user-created tokens. One can broadcast the subject for voting with

---

[197] https://www.tatianamoroz.com/tatiana-coin/
[198] https://techcrunch.com/2014/10/17/bitcoin-2-0-crowdfunding-is-real-crowdfunding/

terms and conditions, and let users vote on its outcome with full transparency by using tokens.
- Instant messenger: GetGems [199] introduces cryptocurrency in a secure, fast Bitcoin Wallet with all of the messaging features of Telegram, the world most encrypted stand-alone instant messenger. GetGems makes handling Bitcoin as simple as sending a text message.
- Distributed computing: FoldingCoin developed a platform to harness home computing power for medical and scientific projects. FoldingCoin rewards the participants with cryptocurrencies based on the computing power contributed. [200]
- Game: The gaming company Everdreamsoft [201] has developed Spells of Genesis as a blockchain based trading card game on the Counterparty platform. Spells of Genesis is a mobile game mixing the collection and strategic aspects of Trading Card games with the addictiveness of Arcade games. While working as a core of the whole game economy and background story, Bitcoin and other cryptocurrencies will become an in-game means of exchange. The gamers can exchange and trade game items easily and freely traded within or without the app.
- RealEstate Coin: RealEstate Coin allows one to invest Bitcoin into U.S. commercial real estate. This app token gives common investors the opportunity to crowdfunding acquisitions of net-leased commercial real estate, and secure payouts from the monthly lease payments of the tenant.

Section 7.11   AML & KYC

---

[199] http://getgems.org/#/
[200] https://foldingcoin.net/index.php
[201] https://www.everdreamsoft.com/

Other blockchain platforms

The blockchain is ideal to provide the integrated decentralized monitoring efforts of financial transactions.[202] Anti-Money Laundering (AML) and Know Your Customer (KYC) practices have a strong potential for being adapted to the blockchain.

Currently, financial institutions must perform a labor-intensive multi-step review/ credit check process for each new customer. Using blockchain can reduce the costs of KYC substantially through cross-institutional client verification using blockchain. At the same time, blockchain can increase monitoring and analysis effectiveness.

Likewise, an anti-money laundering system built on the blockchain can leverage the cryptographically secure, decentralized and immutable nature of the technology to identify and stop suspicious transactions effectively. Smart contracts will allow financial institutions to parse data securely through an AML engine on the blockchain automatically, providing high efficiency and minimum friction.

Each financial institution participating in the system serves as a node within the private permissioned blockchain network and uses the network directory and smart contracts to record transactions on the blockchain.

Since relevant information is stored in the blockchain and made available to each node, all related participants can detect suspicious activity. Anyone detecting suspicious activity can issue a warning to stakeholders and the transaction, and flag and stop the activity for further investigation. This is like to turn all citizens into informers. The blockchain network updates itself immediately with the record of such an alert in an immutable and tamper-proof manner.

---

202 "Blockchain for AML – harnessing blockchain technology to detect and prevent money laundering", Floyd DCosta, https://internationalbanker.com/technology/blockchain-aml-harnessing-blockchain-technology-detect-prevent-money-laundering/

In April 2017, the Japanese government passed a law obligating all Bitcoin exchanges in Japan to implement KYC and AML mandates. Startup Polycoin[203] has an AML/KYC solution that involves analyzing transactions. Those transactions identified as being suspicious are forwarded on to compliance officers. Another startup Tradle[204] is developing an application called Trust in Motion (TiM), which allows customers to take a snapshot of key documents. Once verified by the bank, this data is cryptographically stored on the blockchain.

---

[203] https://polycoin.io/
[204] https://tradle.io/

Other blockchain platforms

Chapter 8 Other blockchain applications

Section 8.01 Data management

In the MDL world, data fall into three categories: identity data, transaction data, and content data. The Identity MDL, Transaction MDL, and Content MDL are three pillars of the data management applications by blockchain technology.

Data contains vital information about persons, organizations, and events. Today, institutions, organizations or companies build their own silos of data and information-management protocols. The protection of these data against unauthorized access or manipulation is important. In 2013, the Target data breach costed Target US$18.5 million to settle. [205] In 2015, hackers obtained personal details, Social Security numbers, fingerprints, employment history, and financial information for about 20 million individuals in a US government database.

Blockchain technology could simplify the management of the trusted information, making it easier for entities/ persons to access and use critical public-sector data while maintaining the

---

[205] "Target will pay $18.5 million in settlement with states over 2013 data breach", Samantha Masunaga, http://www.latimes.com/business/la-fi-target-credit-settlement-20170523-story.html

Other blockchain applications

security of this information.[206] Blockchain technology uses blocks of data. Once these blocks form a chain, they are secure. Verification and management using automation and shared protocols protect the data from unauthorized access.

Blockchain can also help financial companies; government agencies, and others to digitize and manage existing records within a secure infrastructure, allowing agencies to make some of these records "smart." Rules and algorithms allow specific data in a blockchain to be shared with the third party once predefined conditions are met and the third party identity is verified.

Some countries are more aggressive in applying such technology. For example, Estonia is rolling out a platform called Keyless Signature Infrastructure (KSI) to safeguard all public-sector data. KSI creates hashes of the original data. The hashes are stored in a blockchain and distributed across a network of government computers. Whenever underlying file changes, a new hash value appends to the chain. An unauthorized data change will produce a hash not acceptable by the blockchain.

The transparency of the history of each record detects and prevents unauthorized tampering. Government officials can monitor the "who, what and when" of any change. The health records of all Estonian citizens are managed using KSI platform, which is available to all government agencies and private-sector companies in the country.

Besides the protection of data against unauthorized access and change, data sharing is another dilemma in the data management. Until now, the data sharing is a non-reversible process. Once a party shares data, it possesses the data. In the real world, a non-disclosure agreement is a legal instrument to prevent further spread of the data. However, in most circumstances, it is not very practical. For example, if you hand out your credit card information, there is no way to take it back. You must trust the

---

[206] "Using blockchain to improve data management in public sector", Steve Cheng et. al.,
https://www.mckinsey.com/business-functions/digital-mckinsey/our-insights/using-blockchain-to-improve-data-management-in-the-public-sector

other party not to use your information for any unintended purpose.

To solve this problem, the MIT developed a project called Enigma,[207] which is a computation platform to allow data sets to be shared for the computational purpose and yet without having the party using the data to access to the raw data itself. Only the original data owner ever sees the raw data. Enigma protocol is a decentralized, open, secure data marketplace. It opens a new way to share and manage the data and yet still keep data private. In the credit card example above, the merchant submits the transaction to receive payment and yet he does not have your credit card information for any other unauthorized use.

Section 8.02   Distributed data storage & ERP service

Besides managing data, blockchain technology, as decentralizing file storage on the internet, brings clear benefits as well. Today, HTTP downloads a file from a single computer at a time, instead of getting pieces from multiple computers simultaneously. A P2P approach could save 60% in bandwidth costs.

Not only a distributed data throughout the network saves bandwidth, but it also protects files from getting hacked or lost. The data stored in the traditional cloud is not as safe as one might think. For example, Dropbox reset passwords for 68 million accounts in response to a 2012 breach. [208]

On August 31, 2014, a collection of almost 500 private pictures of various celebrities was posted on the imageboard

---

[207] https://www.enigma.co/
[208] "DropBox's 2012 breach was worse than the company first announced", Russell Brandom, https://www.theverge.com/2016/8/31/12727404/dropbox-breach-passwords-hacked-encrypted

Other blockchain applications

4Chan.[209] The images were hacked due to a security issue in the iCloud API via phishing attacks.

Public cloud also has the issue of privacy and potential government surveillance. A proposed solution is to use a private cloud device. Yet it does not address the fundamental problem. It can be more vulnerable than the public cloud. The only advantage of a private cloud is its obscurity. With the arriving of blockchain technology, finally, a solution is on hand to solve such a security issue.

Inter-Planetary File System (IPFS) conceptualizes how a distributed web might operate safely.[210] Similar to the way a BitTorrent moves data around the internet, IPFS removes the need for centralized client-server relationships. Each file is broken into many blocks, which possess unique hashes. The hashes are the index to locate where the blocks of files are stored. IPFS tracks index version history for every file. Each network node stores only one or more blocks of the file. To look up a file, the unique hash index of the file locates the nodes where blocks of the file are stored. The data downloading from the distributed web is no longer sequential, but parallel. It can speed up file transfer and stream greatly without increasing the bandwidth.

Storj provides blockchain based end-to-end encrypted distributed object storage. Only the data owner can access the stored data. It is safe, fast, always available and cheap. It uses blockchains to create a decentralized cloud system using spare disk space allocated by a community of "farmers" who receive rent in Storj's native cryptocurrency.

The software fragments the stored files into small packets, much like the packets on the internet. The blockchain's transaction ledger, public/private key encryption, and cryptographic hash functions encrypt packets and distribute them throughout the network for storage. The blockchain stores information such as the network locations of each packet and its cryptographic hash as

---

[209] "4chan Chronicle/The Australian Hack", Wikibooks, https://en.wikibooks.org/wiki/4chan_Chronicle/The_Australian_Hack

[210] https://ipfs.io/

proof of storage, verifying that the farmer still has that shard and that it is unmodified. Each packet has three copies for redundancy.

The blockchain storage is a tamper-proof way of storing records such as land ownership, business licenses, or birth and death certificates, governments are interested in blockchain-based distributed ledgers. We will discuss this in Section 8.09 of land title registration, identity MDLs.

There is plenty of commercial interest in the file storage solution using blockchain technology. A few companies already use specially encoded Satoshi's – the smallest denomination of Bitcoin – to store ownership ledgers. For example, Everledger tracks diamonds this way to help fight insurance fraud. Similarly, the companies behind the IBM-led Hyperledger believe they can use the blockchain-based, Permissioned technology to track the ownership and exchange of all sorts of things, from stocks and shares to cars and houses.

One of the largest database and ERP companies, Oracle, considers blockchain will fundamentally transform how business is done, making business-to-business interactions more secure, transparent, and efficient. Its impact can be bigger than the social web, Big Data, the cloud, or even artificial intelligence. Oracle is offering the service to enable customers to extend their current Oracle ERP and SaaS solutions to the blockchain-based platform. [211] The major advantages include enabling of the trust in transactions, avoiding the risks of intermediaries, reducing error-prone information exchange processes, avoiding delays of reconciliations, reducing cross-ERP discrepancies, decreasing the cost and improving visibility within the ecosystem.

Section 8.03 Forecasting and prediction

Blockchain technology also finds applications in prediction and forecasting. A company called Augur developed a prediction market platform using the Ethereum blockchain.[212]

---

[211] "Oracle blockchain cloud service"
https://www.oracle.com/cloud/blockchain/index.html
[212] https://augur.net/

Other blockchain applications

Augur's platform is like a betting machine. It allows users to buy and sell shares of the outcome of an event. The share price is the probability of the event actually occurring. The price of each share is one dollar (or one ETH) times the probability. To buy a share to predict an event that it has 60% probability occurring, it costs 60 cents. If it turns out correct, the buyer gets 1 dollar per share back. If not, the buyer loses 60 cents.

As more people buy shares in an outcome, the price of that outcome will rise while the price of the other outcome will fall. The price may rise or fall during the course of prediction since crowd's prediction can change per real-world occurrences. The concept is similar to the binary option, which predicts the stock price, except binary option, does not have enough volume to influence the outcome. The one who initiates the subject of prediction is called the market maker. He can pose any subject for betting, such as the Dow Jones Index three months later. He provides the initial funding and receives trading fees in return.

People have been betting on horse races for a long time. Betting on the outcome of formal horse races can be fun and profitable if you know what you're doing. There are betting on many other markets as well: Auto racing, Baseball, Basketball, Boxing, Football, Golf, Hockey, etc.

With the blockchain technology, the betting is becoming more than betting. Say, if millions are predicting the outcome of an election, or the direction of the stock market, the prediction may become true. Augur combines the magic of betting with the power of a decentralized network to create a forecasting tool.

The crowd predictions have a high degree of accuracy in certain events whose outcome can be influenced by the crowd, but it has no predictability for the events that the participating crowd does not have any control over, such as weather forecast. Augur's prediction markets provide powerful predictive data - you can think of the current market price of any share in any market as an estimate of the probability of that outcome actually occurring in the real world.

Using Augur, anyone, anywhere in the world can create a prediction market quickly and easily. Prediction markets have proven to be more accurate at forecasting the future than individual experts, surveys or traditional opinion polling due to the so-called the wisdom of the crowd.

The collection of the intelligence of many people is superior to the smartest people in the world. They provide real-time predictive data and are traded using real money - which incentivizes market participants to reveal what they think will happen, rather than what they hope will happen. All funds are stored in smart contracts - eliminating counterparty risk and allowing fast, automated payments to winning traders. Blockchain also can automate the deposit and withdraw of funds. With no human intervention required, there is no human error.

Section 8.04   Internet of Things

Like blockchain technology, the Internet of Things (IoT) is another breakthrough technology evolved from the internet. IoT is the core technology of the hot topic of Industrie 4.0. For the first time, the objects are no longer isolated physical entities. The objects, equipped with IoT and sensors, can sense their environment, collect data, communicate with other objects and people, and perform tasks as instructed. The data they collected will be part of the Big Data. Through the two-way communication and Big Data, the network can perform deep learning and develop into artificial intelligence.

IoT promises to bring the fourth industrial revolution since the invention of steam engine, electricity, and the computer. IoT allows people, process, products, manufacturing, usage all integrated together.

IoT's will find widespread adaption in the industries and consumer markets across the board. In 2016, there were already 6.4 billion connected IoT's, up 30% from 2015. By 2025, there will be 75 billion IoT connected devices worldwide according to the recent forecast. [213]

---

213 "Internet of Things (IoT) connected devices installed base worldwide from 2015 to 2025",
https://www.statista.com/statistics/471264/iot-number-of-connected-devices-worldwide/

Other blockchain applications

IoT's, RFID and sensors together form the Wireless Sensor Network (WSN).[214] When an ordinary object attaches such a device, it becomes a part of the WSN (Figure 8-1). It has the ability to infer, see, hear, and measure any environmental parameters (temperature, humidity, chemical composition, etc) through multiple types of attached sensors and actuators. It collects the data as desired and shares the information with the outside world. Since the information transfer is two ways, it also has the capability to receive external instruction to perform functions desired by an external source. In this way, the object is no longer an isolated object somewhere in the world but forms an integral part of a much bigger system.

IoT's is promised to be the next revolutionary technology to integrate not only people but also all the objects into a fully integrated Internet. It can communicate to the internet, by receiving and sending data and information. The exchange of data between objects with IoT and the remote server allows the object to be monitored and controlled. It makes the management of a remote system possible. The result increases system efficiency and improves cost monitoring.

In the industrial applications, IoT on each machine runs the full range of productivity improvements from predictive maintenance of the equipment to the customized configuration of products in the production line. It also provides equipment vendors with invaluable performance data from the field, and the capability of remote diagnosis and troubleshooting to improve their equipment.

By combining blockchain infrastructure with integrated IoT protocol, it further unleashes the power of IoT. The combination of IoT and blockchain will allow any type of transaction to occur between two IoT enabled objects. The industry can create a tamper-proof history of the products from their component supply chain to their field operation, in complex value networks with many stakeholders. The possibilities include connected cars, connected light bulbs, HVAC in the green

---

[214] "Wireless sensor network", https://en.wikipedia.org/wiki/Wireless_sensor_network

building management systems, equipment used in hospitals, tracking devices in container ships, and many others.

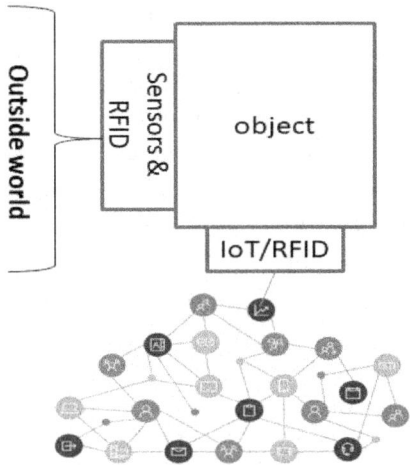

Figure 8-1 An IoT connected object

With the lowering of IoT cost, they will be in the consumer product packages, medical devices, hardware scanners, drones, almost anywhere. The blockchain adds security to the IoT network and allows not only information, data but also valuable assets to be sent over the network.

For example, a proprietary recipe of a pharmaceutical process making a new type of drug developed in the company's R&D laboratory can be sent over to the factory's production IoT platform when released to manufacturing without the risk of being hacked. Alternatively, a highly sensitive technical specification of a sophisticated system in development can be shared among different equipment in different collaboration partners using IoT platform. A smart contract can be sent over the blockchain to a contractor's IoT platform in the supply chain to trigger an event specified in the contract. When a container of goods is loaded on the cargo plane detected by IoT, the smart contract immediately triggers a payment without the involvement of a third party, such as a bank.

There is tremendous potential for the applications of blockchain in industrial use. Samsung and IBM are creating

Other blockchain applications

decentralized networks of the IoT devices using blockchain. IBM's Watson IoT platform enables IoT devices to send data to blockchain ledgers. The platform is called ADEPT (Autonomous Decentralized Peer-to-Peer Telemetry). [215] ADEPT records the transactions carried out by the IoT devices without the need for central control and management. It utilizes hybrid PoW/ PoS to secure transactions. The three building blocks of the platform are BitTorrent for file sharing, Ethereum for smart contracts and TeleHash for peer-to-peer messaging. IoT serves the bridge between the blockchain to the real world.

A Fintech company in China, ZhongAn Technology,[216] subsidiary of ZhongAn Insurance, developed a blockchain-based system that aims to track the whole process of the chicken farming, in partnership with Wopu, also an IoT company in China. The blockchain application will give each chicken an identity by attaching an IoT device to the chicken and uses mobile applications to check each transaction recorded on the blockchain that can trace and record the information of its birth, the farms on which it was raised and the processing factories and logistic suppliers it reaches on its way to market. ZhongAn is also rolling out the first blockchain application in the country's agricultural industry.

Walmart collaborated with Tsinghua University and IBM Blockchain to form WFSCC (Walmart Food Safety Collaboration Center) in Beijing.[217] The project is to use blockchains to explore the traceability of food supply chain and establish the authenticity.

Besides the above examples, there is much more rushing into this new arena, including Fortune 500 companies and startups: Ambisafe, BitSE, Chronicled, ConsenSys, Distributed, Filament, Hashed Health, Ledger, Skuchain, Slock.it, BNY Mellon, Bosch, Cisco, Gemalto, and Foxconn.

---

[215] "IBM reveals Proof of Concept for blockchain powered IoT", Stan Higgins, https://www.coindesk.com/ibm-reveals-proof-concept-blockchain-powered-internet-things/
[216] https://www.zhongan.io/en/aboutus
[217] https://www.walmartfoodsafetychina.com/our-work/policy-support

Leading manufacturing and IT companies, blockchain companies and financial institutions, such as Cisco, Foxconn, Bank of New York Mellon BitSE, Bosch, Gemalto, Chronicled and ConsenSys formed a consortium to develop the Blockchain-based IoT applications. The consortium wants to establish a decentralized and immutable Blockchain protocol as a shared platform to build IoT devices, applications, and networks. The consortium is working to set the shared protocol first. It will then begin to explore structures of IoT networks and design appropriate applications.

There is tremendous potential for the applications of Blockchain in Industrie 4.0. It essentially solves the problem of data security, transparency, and immutability to develop effective solutions for industrial and business operations and processes. In other words, it guarantees the data integrity and creates a tamperproof history of the products through their life cycle with many stakeholders. Each object, physical property or raw material will have its own secured identity.

Not only manufacturing industry, service industries such as healthcare can also use IoT plus blockchain to drastically improve its service by attaching such devices to the patient. Multiple parties can gain real-time access to information being distributed and recorded onto a public Blockchain, such as the monitoring of insulin level or heart rate. The potential of the combined IoT and blockchain is beyond imagination.

Section 8.05 IOTA and IoT Chain

IOTA is designed to close the gap between the world of IoT and the blockchain. IOTA[218,219] is the brainchild of the IOTA Foundation - a German company. It envisions that by 2025, there will be more than 70 billion IoT connected objects worldwide. Each IoT enabled object will generate data. Data are valuable and sharing the data will create explosive growth in the knowledge-based economy. Whenever there is an exchange of valuables, there is a market.

---

[218] http://iota.org/
[219] http://iotatoken.com/IOTA_Whitepaper.pdf

Other blockchain applications

IOTA is a public Permissionless distributed ledger using a new data structure, called Tangle, to replace the block. Tangle is based on the Directed Acyclic Graph (DAG). In mathematics, DAG is a topological ordering that is in one direction only and has no return.[220]

Tangle has no blocks, no chains, and no miners.[221] Without the need to create blocks, there is no fee for the block creators. Therefore, the Tangle data structure is a truly feeless distributed ledger. Without blockchain, the transactions are linked in such a way that each participant must have approved two past transactions to make a new transaction. This is shown in Figure 8-2. The very first transactions are called genesis transactions. They are created by the tokens.

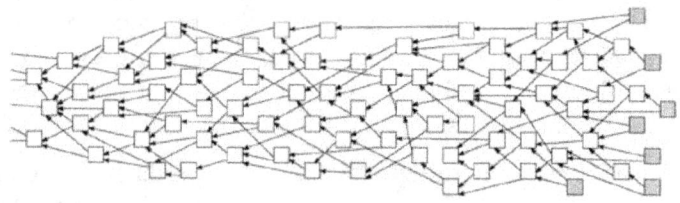

Figure 8-2 Tangle data structure (source: https://iota.readme.io/ docs/what-is-iota)

On the bitcoin network, many transactions are mined into blocks. However, in DAG, one can consider each transaction is a block. Every transaction directly maintains the sequence. There is no need for the process of mining. This makes it more efficient. On the other hand, on the DAG network, each validated transaction needs to link to an existing transaction just like the block links to the chain. However, since there many parallel links, it would make the network grow horizontally and become very wide. IOTA proposed its own algorithm controlling the width of the tangle network.

IoT Chain (ITC) is another distributed ledger built on DAG for the IoT applications. It can handle over 10,000 transactions per

---

[220] https://en.wikipedia.org/wiki/Directed_acyclic_graph
[221] https://iota.readme.io/docs/what-is-iota

second. What is unique about ITC is that it can run raspberry pi-level low-performance IoT devices. It is essentially an IoT operating system using DAG. It also has tokens for the settlement of smart devices using rights, ownership transfers, and other value transfers in the entire IoT ecosystem. IoT Chain has the potential of becoming a new variation of the blockchain. DAG can process thousands of transactions per second; therefore, it can be used for large-scale applications. It beats Ethereum's solution of scaling called sharding, a database partitioning technique to divide a very large database into smaller components easy to manage.

Today, according to the estimate that more than 95% of the data are not shared because the owner of the data does not have any economic incentive to share them. The most noticeable shared data are intellectual properties, such as movies, music, books, which have a long history of having the market on its own. There are some blockchain applications already aiming at such a market. We will discuss more in Section 8.11. IOTA wants to expand the data market to include any valuable data and be the cryptocurrency for such a market. IOTA Foundation called it the "machine economy" and "data economy".[222]

The data center in the early 21st century is akin to the power plant in the early 20th century. Then, the large-scale power plants sprang into existence to bring electricity to each home. Today, the data center brings data to each person through the internet. With the coins to trade data, it could unleash a huge amount of locked data for the widespread use. This will have an unimaginable and profound impact on everything as we know of today. No wonder, IOTA market value has leaped to the 4th place of $12 billion dollars in relatively short time.

IOTA is collaborating with Taipei city to make Taipei, Taiwan into a smart-city.[223] Upon the completion of the project, it will turn Taiwan's capital into a functional and operational smart city. This includes an ID built-in, which will prevent identity theft

---

[222] "IOTA data market place", David Sonstebo,
https://blog.iota.org/
[223] https://globalcoinreport.com/iota-miota-is-trading-up-while-testing-digital-ids/

Other blockchain applications

and every other criminal identity related to manipulating with identification documents.

In December 2017, IOAT has also entered a partnership with world-renowned companies such as Samsung, Microsoft, Cisco, Fujitsu, and Volkswagen to develop the data market.

Section 8.06   Supply chain management

In the supply chain management, MDL can offer many breakthrough applications, such as product authentication and innovative supply chain financing.

In product authentication application, MDL is the ideal platform to certify the authenticity, fair trade status and origin of the products, components or raw materials in a supply chain. Transparency comes with blockchain-based time stamping of a date and location — on ethical diamonds, for instance — that corresponds to a product number. It also reduces the cost, labor, time in managing the supply chain.

In this arena, the UK-based Provenance[224] offers supply chain audition for a range of consumer goods. Making use of the Ethereum blockchain, a Provenance pilot project ensures that the suppliers in Indonesia have harvested fish sold in Sushi restaurants in Japan in a sustainable way.

For the supply chain financing, companies like Skuchain of Mountain View in California,[225] Hijro and Fluent in Lexington, Kentucky, are developing financing services for the buyers and sellers.

Since the interlocked interests of the parties in a supply chain are across multiple businesses, supply chain financing can be more creative than the regular financing. Supply chain financing provides an invaluable opportunity to strengthen relations with suppliers. The objectives of supply chain financing are more than just to earn interest, but to ensure the deal goes through.

---

[224] https://www.provenance.org/
[225] http://www.skuchain.com/

Supply chain financing by blockchain can optimize the utilization of working capital within the supply chain, creating mutually beneficial arrangements between buyers and suppliers. The smart contract can facilitate lending by financiers against purchase orders, invoices, inventory assets and payment obligations. The automated release of funds triggered by real-world events offers the assurance to the loans. It provides a real-time, reliable view of the transaction state, bringing significant transparency for all participants and helping them build a more trustworthy and stable supply chain ecosystem. It also enhances the liquidity of collateralized assets in a supply chain by improving upon current trade finance instruments such as Factoring, PO Financing, and Vendor Managed Inventory Financing.

Services like IMT offered by Skuchain provide inventory financing. The idea is to assign the original purchase contract between the buyer and the seller on the ITM blockchain. The contract provides the collateral to an investor in the IMT fund. IMT uses its funds to purchase goods from the seller. Each purchase order triggers the shipment of the finished goods according to the order. The buyer then pays IMT for the goods.

The platform developed by Fluent a Bitcoin-like system for the supply chain financing.[226] Buyers approve invoices on the Fluent Network once they receive the goods satisfactorily. The Fluent Network's blockchain is operating on top of the existing banking infrastructure. The actual funds remain in custody of banks at all times. On the blockchain, the funds are converted into tokens for the transactions.

Not only the tech startups, many financial institutions, as well as global enterprises, which are already part of the existing supply chain network, are developing blockchain based supply chain applications. A single blockchain platform with

---

[226] "How Fluent Wants to Streamline Financial Supply Chains With a Blockchain", Aaron Van Wirdum, https://bitcoinmagazine.com/articles/how-fluent-wants-to-streamline-financial-supply-chains-with-a-blockchain-1465318410/

cryptographically verified invoices, instant settlement, and low operating costs benefits all parties on the system.

Foxconn, the world's largest contract manufacturing company, which makes all the iPhones in the world, has launched a new blockchain-based supply chain finance platform, called Chained Finance based in Shanghai, through its financial service arm – FnConn.[227] It collaborates with Dianrong.com,[228] the first Chinese P2P company to participate in Hyperledger Blockchain Project. The new platform leverages advanced financial technology to meet the financial need of the SME's involved in the supply chain in China.[229] SME's are making up most of the supply chain, but they have limited financing options. Using dynamic discounting technique can enable buyers and suppliers to negotiate payment terms and discounts; when buyers have capital available that can be redeployed to suppliers in exchange for better trading terms. The smart contract can implement such dynamic discounting scheme easily. Chained Finance enables to deliver needed capital to these suppliers and provides large multinational manufacturers with enhanced visibility and transparency.

Chained Finance is initially targeting three major industries: electronics, auto, and garment manufacturing. The blockchain is revolutionizing the finance industry. It can offer solutions to any company operating and financing complicated supply chains. Chained Finance creates a unique ecosystem that will provide supply chains with easier access to funding at competitive rates. In return, supply chain operators will gain greater visibility of their suppliers and the many layers of finance embedded in the process.

---

[227] "Foxconn reveals plan for blockchain supply chain domination", Michael del Castillo, https://www.coindesk.com/foxconn-wants-take-global-supply-chain-blockchain/

[228] http://en.dianrong.com/

[229] "Chained Finance: First Blockchain Platform for Supply Chain Finance", https://www.prnewswire.com/news-releases/chained-finance-first-blockchain-platform-for-supply-chain-finance-300418265.html

By using the Chained Finance platform, every payment, every supply chain transaction, can be more transparent, manageable and easily authenticated. Chained Finance will help eliminate many of the trust issues faced by counterparties and deliver automated execution. The new platform will be an enabler of supply chains across many major industries and geographies.

Chained Finance has an initial focus on the automotive, electronics and garment production industries, though it has much wider applications. By using the Chained Finance platform, every payment, every supply chain transaction, can be more transparent, manageable and easily authenticated. Chained Finance will provide timely, efficient support to far more suppliers of all sizes. It will also help ensure the timely delivery of products to end customers and improve efficiencies across the entire supply chain.

Founded in 2012, Dianrong is now a leader in online marketplace lending in China, dispersing more than US$300 million loan amount a month to 3.7 million retail lenders. Its products and services include loan originations, investment products, and marketplace lending solutions in a comprehensive, one-stop financial platform supported by industry-leading technology, compliance, and transparency. The company's sophisticated and flexible infrastructure enables it to design and customize lending and borrowing products and services, based on industry-specific data and insights, all supported by online risk-management and operation tools.

In summary, the blockchain will revolutionize the supply chain system to make it more efficient, less costly and more trustworthy.

Section 8.07 Governance and voting

A public blockchain allows people who do not know each other to agree to a set of rules for the governance purpose. For example, blockchain can verify that no one has tampered with the file, which stores digital signatures and provides identity verification. This could allow citizens to self-service many bureaucratic transactions that require the attention of civil servants and lawyers today.

Other blockchain applications

Different public blockchains have different rules and incentives, but fundamentally, they all utilize rules and incentives to bring people together to perform certain functions. This is a definition of the governance. Best of all, with blockchain, it can be done without relying on the centralized organization to reach consensus. Therefore, blockchains are ideally suitable for governance.

By making the results fully transparent and publicly accessible, MDL technology could bring full transparency to elections or any other kind of poll taking. Ethereum-based smart contracts help to automate the process. As such, there are efforts to develop blockchain technology for the governance purpose that potentially is able to redesign our interactions in business, politics, and society.

Blockchain platforms can manage social interactions and governing rules. They can be useful and important tools for the responsible party in the governance, but they cannot replace traditional central authorities, which have the role to define the rules. Thus, the blockchain apps for governance are mostly to be a Permissioned blockchain. The ownership must belong to a responsible governing entity rather than an anonymous public.

For example, the app, Boardroom,[230] enables organizational decision-making to happen on the blockchain. The Boardroom is a Governance DApp for individuals and companies to manage their smart contract systems on the Permissioned Ethereum blockchain. It can conduct proxy voting for shareholders on the board proposal. It is an administrative tool for organizations. These are very narrowly defined organizations, simply concerned with bookkeeping of address balances, but they meet all the requirements of decentralized decision-making.

Decentralized autonomous organizations (DAOs) are virtual entities comprised of a large number of individuals that respond to a well-defined set of rules. A social structure that constitutes 'governance' is the one that organizes the rules. The economic and governance structures that most suitable for the blockchain governance application are those which motivate the resulting consensus protocol properties.

---

[230] http://boardroom.to/

Separating corporate governance from the underlying consensus process entirely will advance the longevity and integrity of DACs by improving their transparency. The voting process is particularly suitable to employ blockchain technology to make it tamper-proof.

Today, the voting systems are either a paper-based system or a digital voting system. The security risk of digital voting is even higher than that of the paper voting system because the digital voting system is more prone to fraud than the paper voting system.

The digital voting system identifies a voter through his/ her ID card. The platform must confirm a voter's identity before he/ she can cast their vote in the polling station. When a voter submits their vote, it goes to a vote storage server. The vote storage server encrypts the vote, and strips voter's ID data from the vote, before sending it to a vote-counting server. The vote-counting server decrypts and counts the votes and then outputs the results. In today's digital voting platform, there is a risk of an attack before the votes are transferred to the counting server.

Blockchain technology can develop a secure and robust voting system. The blockchain does not need to replace the current voting system but rather to improve it. The blockchain voting system first registers the voter into the so-called the registration blockchain. The registration creates a transaction unique to the voter. The government miner validates the user. Upon validation, the voter will receive a ballot card with his/ her information on it. This is the same process as the voter registration in the US today, except that the registration is a record in a blockchain.

The voting itself actually takes place in a separate blockchain – the voting blockchain. Each vote is a transaction. The platform validates a voter by three-factor authentication; their identification number, the password supplied on registration, their ballot card that contains a QR code. The voter can only vote in his/ her constituency where he/ she registered.

The separation of registration and voting blockchains ensure voter anonymity. Certified observers and election officials will monitor and audit the voting process. Instead of reading the ballots and counting them, they will host the nodes in the voting

Other blockchain applications

blockchain and verify that the unencrypted results match the encrypted votes. Each transaction is encrypted with public and private keys. The polling station nodes contain the public keys of their voters to allow them to encrypt any vote made to that polling station.

Each constituency counts the votes and builds blocks. They act as miners during the counting process. They build the blocks by including the votes for the same candidate in a block. A block can contain five thousands or ten thousands of votes. This is happening in the local constituencies all over the country. The local polling stations then broadcast their voting blocks to all nodes of voting blockchain network of the same constituency to build the blockchain. Once the vote has been confirmed or built into a block, the polling station will then generate a transaction to remove the vote in its pool memory. Such a voting system is tamper-proof.

Besides the voting system, blockchain technology can benefit the bureaucratic government in many other applications. Governments are known for their inefficiency because of their size, inertia, bureaucracy, and lack of incentives. The services are normally slow negatively impacting citizens. Linking the data between the departments with blockchain ensures the real-time release of critical data. For example, the real-time consumer information can help to make economic and monetary policy. A real-time nationwide clinical data can help to contain the spread of a contagious disease.

Blockchain technology could improve transparency and check corruption in governments worldwide. For example, the US Navy's Innovation department recently announced its interest in using blockchain technology for their manufacturing system. They plan to add blockchain technology to their 3D printing in order to help securely transfer data through the manufacturing process.[231] The Naval Innovation Advisory Council will test the blockchain technology integration into their system. The initial test is to prove

---

[231] "The US Navy wants to connect its 3D printers with a blockchain", Stan Higgins, https://www.coindesk.com/the-us-navy-wants-to-connect-its-3-d-printers-with-a-blockchain/

the concepts, share data, and secure digital designs throughout the Navy's network of information.

They also plan to create a data-sharing layer among the 3D printing sites using Blockchain technology. Other government departments, such as Department of Homeland Security Science and Technology Directorate, awarded a total sum of $9.7 million to several small businesses for technology research on Blockchain usage, such as contract bidding.[232]

Section 8.08   Regulatory applications

Government regulation is possibly one of the most important factors as to whether or how the blockchain technology will eventually develop into a full-fledged Fintech industry. RegTech (Regulatory Technology) is a branch of Fintech.[233] It specifically applies to the regulation related challenges of Fintech, including fraud detection and prevention. There are challenges to understand, implement, embed and enforce the new legislation and regulation in the Fintech and blockchain systems. RegTech explores how firms can benefit, leverage, better understand and manage the risks to comply the new legislation and regulation. Since the Fintech and blockchain technology are still evolving quickly, so as the related legislation and regulations, the situation is fluid.

RegTech companies have expert knowledge of both technology and regulation. RegTech provides executives with the tool to introduce new capabilities to leverage systems based on the new legislation and regulation. It also allows them to analyze regulatory data and report in a cost-effective, flexible and timely manner, be able to respond and react as new regulations emerged. RegTech can assist industry in complying with regulation and on the other hand, regulators can use RegTech to make better use of

---

[232] "Blockchain applications for Homeland Security analytics", https://www.sbir.gov/sbirsearch/detail/867813

[233] "Is RegTech the new Fintech?", https://www2.deloitte.com/ie/en/pages/financial-services/articles/RegTech-is-the-new-FinTech.html

Other blockchain applications

the information provided by the industry to update the legislation and regulation.

The scope of RegTech is big: It needs to cover everything from systems that monitor and control core ledgers to the "purses" on the periphery that store value locally with users. Regulators could insist on people recording transactions externally on MDLs, providing open sources of transaction prices and volumes, or increasing competition through increased data portability.

RegTech solutions are mostly Cloud-based with remote mining, maintenance management, and backup of data. RegTech apps provide compliance obligation analysis, risk identification, irregularity detection and prevention and management tools for legislation/regulation gap analysis, compliance, management information, transaction reporting, regulatory reporting, activity monitoring, training, risk data warehouses, case management etc. RegTech can help firms to automate the more mundane compliance tasks and reduce operational risks associated with meeting compliance and reporting obligations. It can also empower compliance functions to make informed risk choices and how it mitigates and manages those risks. RegTech is more than the use of technology to meet regulatory requirements. It also uses technology to query legislation, regulations in order to identify compliance imperatives.

Companies, such as Hadoop,[234] Tableau, and Pentaho[235], develop tools to organize data and create a report to meet regulatory requirements. In addition, these tools apply analytics to Big Data. For example, Tableau is a visualization tool that makes it easy to look at your data in new ways to help identify trends and from a regulatory perspective. Traditional financial institutions are also investing in regulatory Fintech: HSBC created a US$200 million fund to heighten levels of regulation in the Fintech sector, aiming at improving new compliance demands.

Examples of RegTech companies include FundRecs,[236] which developed reconciliation software for the Funds Industry;

---

[234] http://hadoop.apache.org/
[235] http://www.pentaho.com/
[236] https://www.fundrecs.com/

Silverfinch,[237] which creates connectivity between asset managers and insurers through a fund data utility in a secure and controlled environment; TransUnion,[238] which prevents online fraud by scanning transactions in real time, FundApps,[239] which offers tools for compliance monitoring and reporting.

One of the challenges of the RegTech development is that the legislation and regulation are different from country to country. It creates both technical and non-technical difficulties to develop a universal RegTech app. While Europe has a European Banking Authority that guides the region, the US has a different standard. As long as there is no standardized regulatory baseline or data interchange format across the regions, there are certain difficulties to develop a universal app. This is a significant opportunity for RegTech solutions to be at the heart of these standardization data mazes.

Section 8.09 Land title registration and real estate

A number of countries are undertaking blockchain-based land registry projects. Honduras announced such an initiative in 2015. In 2016, the Republic of Georgia contracted the Bitfury Group to develop a blockchain system for property titles. More recently, Sweden also announced it was experimenting with a blockchain application for property titles.

Probably Japan put in the biggest effort in such blockchain applications. The lack of sufficient and updated data has been a major problem for the Japanese authorities since a lot of ownership cases cannot be properly solved because of insufficient data. The Japanese government wants to upgrade their real-estate registration systems, using blockchain technology in order to more effectively collect, manage and update real-estate data, and to make the data tamper-proof. A blockchain-based registry would allow authorities to better upkeep real-estate property transactions and boost recovery efforts of properties, in the event of a natural

---

[237] https://www.silverfinch.com/
[238] https://www.transunion.com/idvision
[239] https://www.fundapps.co/

Other blockchain applications

disaster like the famous tsunami of 2011. The Japanese land ministry will launch a trial version of the new system as soon as in the summer of 2018. The government will select a few cities for testing. Japanese government is planning to roll out the system nationwide within 5 years after proving the system feasibility.

Likewise, the Ukrainian government also plans to implement Blockchain solutions to solve problems of a depressed land market due to the lack of suitable financial instruments for leases and land transfers. The lease price is low because of a profound black market. The application of a Blockchain system will be able to protect the auctions from black market controls and therefore to stabilize the land price slide and increase income for farmers. Their pilot project is to transfer the State Land Cadastre to blockchain technology.

These are examples of the blockchain applications in the governmental administration of the land ownership. At the same time, blockchain technology also finds its use in the private real estate business. Real estate transactions have always been cumbersome and complicated. Property titles are a case in point. They tend to be susceptible to fraud, as well as costly and labor-intensive to administer. Blockchain technology can have a great impact on the financial verification of the sales process itself. As publicly-accessible ledgers, blockchains can make all kinds of record-keeping more efficient.

The most costly and complicated process is the use of escrow and title companies for third-party verification, which is, of course, necessary to prevent the risk of fraud. With blockchain, there will no need for the escrow company.

By using a blockchain-distributed database to prove authenticity, homeowners could legitimately transfer ownership immediately without the need to pay for third-party verification.

Blockchain technology can be a help in the rental business as well. The blockchain would effectively make forged ownership documents and false listings outdated, making selling or advertising properties you do not own almost impossible.

Blockchain technology also improves the mortgage business, which is always slow and plagued with red tape and administrative issues. Using the blockchain, buyer, seller, as well as the real estate asset, can all have digital ID's. The mortgage process and transfer of ownership would be seamless and much faster. To verify the credit history and income of buyer would be instantly avoiding time-consuming trips to banks, lawyers, and estate agents. Homeowners would be able to prove ownership of their property with a record of their purchase transaction. Houses could acquire digital identities as well, including the chain of ownership, a list of repairs and renovations and the history of real estate tax payment. When this is implemented, the real estate transaction can happen not in one or two months, but in one week. This will have a profound impact on the real estate market.

Section 8.10   Law and blockchain

Laws are actually contracts between citizens and the government. Blockchain smart contract can implement many civil laws. For example, DMV can revoke a driver license using smart contract for three consecutive traffic offenses automatically. The fines of traffic tickets can be automatically deducted from offenders' bank account. A restaurant's license can be voided if it does not pass the food safety inspection. The smart contracts will have a profound impact on industries. Smart contracts eliminate the intermediary, such as a legal firm, as payment will happen based on meeting certain milestones. By its very nature, the smart contract is easily enforceable electronically, creating a powerful escrow by taking it out of the control of a single party.

Blockchain smart contract will not replace lawyers anytime soon, but they can greatly help lawyers to do their jobs more efficiently. In 2017, ten law firms and four legal institutions including Cooley, Debevoise & Plimpton, Hogan Lovells, and others have jointly formed the Ethereum Enterprise Alliance (EEA).[240] Their objective is to create the framework of legally binding smart contracts.

---

[240] "Legally Binding Smart Contracts? 10 Law Firms Join Enterprise Ethereum Alliance", Michael Del Castillo, https://www.coindesk.com/legally-binding-smart-contracts-9-law-firms-join-enterprise-ethereum-alliance/

Other blockchain applications

EEA is not the only entity interested in the application of blockchain in the legal field. A large law firm, Frost Brown Todd (FBT) has also taken the initiative to deploy smart contracts in the legal field.[241] FBT developed a prototype smart contract to be used in software escrow agreements.

The smart contract, once written, will execute on its own. As the lawyers are not skillful in programming the smart contract, and the programmers are not skillful in legal terms and conditions, the gap may cause undesirable consequences. However, the trend will prevail that smart contracts will bring developers and attorneys together to collaborate and provide progressive solutions for the legal industry.

In addition to the blockchain, cloud service also enters judicial applications. Ali Cloud is one of China's top providers of cloud computing services, domain services, emails, network security and Big Data analysis. With its data and emails preserved for judicial departments as courtroom evidence, this may be the first instance of blockchain being utilized by a state judiciary.

Section 8.11 Protection of intellectual property

Computers can easily reproduce digital contents, such as movies, music, and books, and widely distribute them, causing rampant pirating of illegal copies of the intellectual properties. The unauthorized copies of digital intellectual properties deprive the financial gains of their creators. Smart contracts can protect copyright and automate the sale of creative works online, eliminating the risk of file copying and redistribution. Blockchain protects IP in two ways: registering the content to prove the ownership and collecting the payments when the content is being used.

Blockchain tracks the rights and transactions attached to any types of digital contents from music, books to videos.

---

241 "How the Legal Industry is Adopting Ethereum-based Smart Contracts", https://cointelegraph.com/news/how-the-legal-industry-is-adopting-ethereum-based-smart-contracts

Blockchain technology registers IP rights, catalogs, stores the original works, and makes the content widely available. By registering the IP rights to a blockchain, authors possess tamper-proof evidence of ownership. This is because a blockchain transaction is immutable, so once a work has been registered to a blockchain, that information cannot ever be lost or changed.

Ethereum platform uses blockchain to issue smart contracts to verify contractual agreements. There are now startups using Ethereum platform to focus on IP solutions and other alternative blockchain applications. Mycelia uses the blockchain to create a peer-to-peer music distribution system.[242] Founded by the UK singer-songwriter Imogen Heap, Mycelia enables musicians to sell songs directly to audiences, as well as license samples to producers and share royalties to songwriters and musicians through automated smart contracts. By doing so, it empowers a fair, sustainable and vibrant music industry ecosystem involving all online music interaction services.

There are many other startups to do such work. Ascribe, a digital IP platform offers solutions to solve the issues surrounding IP [243] and digital content on the internet. InterPlanetary File System (IPFS), another startup company, developed a peer-to-peer protocol to do the same.[244]

The blockchain and timestamping services can be used to create an auditable trail of content ownership from creation through to the transfer of rights and beyond. Platforms such as Blockai[245] and Ascribe are allowing authors to make a record of copyright ownership, which can then be used to see where and

---

[242] http://myceliaformusic.org/
[243] "How Ascribe uses Bitcoin tech to help underserved artists", Stan Higgins, https://www.coindesk.com/ascribe-bitcoin-tech-underserved-artists/
[244] "An introduction to IPFS", https://medium.com/@ConsenSys/an-introduction-to-ipfs-9bba4860abd0
[245] "Blockai taps the Bitcoin blockchain to protect creative content", Martin Hsu, https://www.ccn.com/blockai-taps-bitcoin-blockchain-protect-creative-content/

Other blockchain applications

how the work is being used on the internet and to seek licenses from third parties.

Registering a work in a blockchain gives the author a digital certificate of authenticity, which can help third parties identify the author of a work. The blockchain, being a permanent immutable record, is the perfect solution for providing proof of creation. There is now a universal database for the record of ownership of IP content, and making the payment of its use indisputable.

Once a work is registered and verified in the blockchain platform, authors can make the IP content unusable for those who are not paying the right to use, or to be notified to see who is using their work. Blockchain registration reduces the cost of transactions and creating a direct link between authors and users.

No less important is that blockchain can help is to collect the payments from IP users. Smart contracts assist in the sale and licensing of intellectual property through the micropayments. Each time, a potential user using the content makes a small payment to the author. As a result, the author can be remunerated each time when his/ her IP is being used without having to pay broker fees. Such a method is simpler and more transparent than many other existing means of payment for authors.

Ujo Music,[246] an Ethereum based and ConsenSys backed music software services company, has used its blockchain application with singer-songwriter Imogen Heap to release her songs on the blockchain. Users are able to purchase licenses to download, stream, remix and sync the song via smart contracts, with each payment automatically split on the blockchain and sent to Imogen Heap directly.

Section 8.12   Healthcare

One of the most important infrastructures of the healthcare industry is the data management. Using digital signatures on blockchain-based data allows access to the health records only by authorized people. A community of people, including hospitals,

---

[246] https://blog.ujomusic.com/

doctors, patients, and insurance companies, could be part of the overall blockchain, reducing fraud in healthcare payments.

A blockchain-based system could ensure that care allowance is spent exclusively on healthcare activities. The system can save time spent on reconciliation after every transaction, helping with straight-through processing.

IBM has teamed up with the Centers for Disease Control and Prevention (CDC) as well as US Food and Drug Administration (FDA) to develop blockchain and distributed ledger technology (DLT) applications in the health sector.[247]

The project with FDA is to define a secure, efficient and scalable exchange of health data using blockchain technology, with an initial focus on oncology-related data. IBM and the FDA are exploring the exchange of owner-mediated data from several sources, such as electronic medical records, clinical trials, genomic data and health data from mobile devices, wearables and the Internet of Things (IoT).

In 2016, IBM and the New York Genome Center jointly create a comprehensive and open repository of genetic data to accelerate cancer research and scale access to precision medicine using Watson, IBM's artificial intelligence (AI) system.

The CDC is working on several proofs of concept based on blockchain technology. It plans to build real DLT applications for the public health sector[248]. The CDC, state and local health departments and other organizations routinely share public health data so they can control the spread of a range of infectious diseases. Currently, the task is managed by the traditional database. Such a database is extremely complicated because the vast areas, organizations, nature of data are involved. Blockchain technology could automate many of the database management processes, store and share health data faster, more secure and incorruptible. Patient's data can have much better privacy and

---

[247] "IBM partners with CDC to bring blockchains to public health" by Giuliu Prisco, http://distributed.com
[248] "Why CDC wants in on Blockchain" by Mike Orcutt, MIT Technology Review.

Other blockchain applications

security. Critical data can be trusted. This can be especially important when managing a public health crisis.

In addition, by combining the blockchain and Artificial Intelligence (AI) technologies, while DLT securely manages and shares data and AI extracts embedded patterns from the data. It opens up a vast frontier for exploration.

The amount of research and clinical data in the healthcare sector is exploding. Blockchain technology is coming at the right time to enable the ecosystem of data in healthcare to have more fluidity.

Section 8.13   Food safety

Food poisoning is one of the major health issues. Every year, 400,000 people die due to contaminated food worldwide. In the US alone, CDC estimates 48 million people get sick, 128,000 are hospitalized and 3,000 die from food-borne diseases each year.[249] In December 2006, E. Coli outbreak affected 71 customers in Taco Bell across five states. Eight people developed kidney failure, and 53 people were hospitalized. From October to November 2015, Chipotle Mexican Grill E. Coli outbreak affected 55 customers. [250] In 2009, the Peanut Corporation of America (PCA) experienced a Salmonella outbreak. 714 people got sick and nine died. The company went bankrupt.

Food contamination can happen due to many reasons: cross-contamination, the spread of food-borne illness, unsafe practices, etc. From outbreak to the identification of the source of contamination can take weeks or longer, causing problems to spread. Blockchain technology can help to trace the contamination much quicker, therefore, limit the scope of the damage, both in terms of the people affected and the lost business. IBM entered collaboration with major food suppliers, such as Dole, Walmart,

---

[249] CDC, https://www.cdc.gov/foodborneburden/index.html
[250] "Worst Food borne Illness Outbreaks in Recent U.S. History", https://www.healthline.com/health/worst-foodborne-illness-outbreaks

Golden State Foods, Kroger and many others to address food safety issues worldwide using the blockchain technology. [251]

Blockchain technology can improve food traceability by providing trusted information on the origin and state of food. All participants in the food supply chain, from growers to distributors all way to consumers, can gain access to the known and trusted data of the origin and state of food for their transactions. Such chained information greatly shortens the tracing time.

The food supply members of the consortium identify and prioritize deficiencies in the current food tracing system, while IBM provides technical solutions. Once developed, IBM's fully integrated, enterprise-grade production blockchain platform, which runs on IBM Cloud, can be adopted beyond food supply chain applications. [252] The platform includes new features developed in collaboration with the Hyperledger community, including the Hyperledger Fabric and Hyperledger Composer hosted by the Linux Foundation.

Recently, IBM and Walmart conducted platform trials in both China and US. The trial was a success - the platform can track a product throughout the supply chain's every stage in seconds instead of days or weeks needed in the traditional system. Blockchain technology enables a new era of end-to-end transparency in the global food system.

Section 8.14 Defense industry

Like any other industry, aerospace and defense companies already use a variety of technologies to address operation and product development to improve visibility and efficiency. Yet, due to the nature of their customers and products, which are critical to the national security, there are extra burden and requirements to be

---

[251] "IBM announces major blockchain collaboration with food suppliers", http://www-03.ibm.com/press/us/en/pressrelease/53013.wss

[252] "IBM deploys blockchain technology to provide enterprise solutions to food safety", Brigid McDermott

Other blockchain applications

met that are not required by other industries, such as consumer or agriculture.

Blockchain offers them additional values in maintaining classified information along with their supply chains and critical defense-related infrastructures, such as to track and audit transactions across multiple supply chain and operational partners.[253]

Defense industry infrastructure is dispersed across different locations. Some of these locations serve both defense and non-defense markets. Blockchain technology can be used to ensure security against any unauthorized access to these important network and hardware equipment by consensus-based access.

For example, blockchain can do the certification of people, partners, and parts, e.g. a key component used in F-35 jet requires certification of the part origin. The blockchain-based system can manage the supply chain of hundreds of thousands of components going into a highly sophisticated defense product by recording each transaction step of each component from raw material until finished parts. It can also serve as security clearance to allow certain employees, customers and partners to have the privilege to use certain facilities and tools, or to access certain data. It eliminates the possibility that a back door or a bug was implanted in a critical electronic component, which becomes a key component in the operation of the final product.

Blockchain technology can find applications in both the defense industry and in the military operations. The blockchain is an ideal platform to deliver hack-proof messages. The blockchain platform based system can send and receive smart documents and contracts securely.[254, 255] Crypto-Chat, a subsidiary of Indiana

---

[253] "Blockchain in aerospace and defense", Accenture consulting, https://www.accenture.com/t20170928T023222Z__w__/us-en/_acnmedia/PDF-61/Accenture-Blockchain-For-Aerospace-Defense-PoV-v2.pdf#zoom=50

254 "U.S. Defense finding use cases for Blockchain technology?", J R Cornel, https://steemit.com/blockchain/@jrcornel/u-s-defense-finding-use-cases-for-blockchain-technology

Technology and Manufacturing Companies (ITAMCO), developed a secure messaging and transaction platform for the U.S. military.[256]

Section 8.15   Golem – a world computer

Ethereum network consists of millions of the computer nodes. Many of these computer nodes have idle CPU power ready to be harnessed. On 11 November 2016, the Ethereum community funded a project called "Golem". Golem is an Ethereum plug-in, which enables users and app providers to rent out their unused CPU power and software. It is like Airbnb of CPU power and apps. Any user ranging from a single PC owner to a big data center can contribute resources to Golem network.

Golem is a very special type of blockchain platform. Its aim is to harness unused computing power and storage capacity in the network. Golem is a global, open sourced, decentralized supercomputer that anyone can access. It utilizes the combined power of user's machines, from personal laptops to entire datacenters. There is no single point of failure and no trusted authority. The Golem Network is a decentralized sharing economy of computing power, where anyone can make money 'renting' out his or her computing power or developing & selling software.

Computations in Golem nodes take place inside isolated virtual machines for maximum security. Providers decide how many CPU cores, and how much RAM and disk space they wish to rent to the Golem Network. A reputation system enforces desired behavior of nodes in a decentralized environment, without relying on any supervising institution. This allows nodes to attribute a reputation rank to their peers. A node will get a lower

---

[255] "ITAMCO to Develop Blockchain-Based Secure Messaging App for U.S. Military", https://www.prnewswire.com/news-releases/itamco-to-develop-blockchain-based-secure-messaging-app-for-us-military-300464063.html
[256] http://www.crypto-chat.com/

Other blockchain applications

rank in case of inappropriate behavior and is increased after successful computation.

Developers can create and distribute apps on the Golem Network's Application Registry (Golem Store) and choose whether they want to charge for their apps. Thus, Golem Network has CPU power, storage space, and apps, all the necessary elements of a computer. In fact, it is more powerful than the fastest supercomputer in existence. According to some estimate, Golem can do 2 trillion operations per second. With such a supercomputer available at one's fingertip, tasks such as scientific research, graphics rendering, artificial intelligence, machine learning and data analysis all can be done a fraction of the cost and faster than today's supercomputer.

Many advanced technologies have complementary capabilities, such as IPFS/Filecoin[257] (distributed file sharing and storage service), Whisper[258] (an app securely broadcast messages between users), DEVp2p[259] (a cross-platform peer-to-peer client library, for desktops and mobiles), Swarm (a BitTorrent-like protocol to enable the transfer of actual files), Ethereum and many more to come. By deploying them together, the combined platform can replace the huge data centers that currently power the internet, and become the decentralized computing power behind the entire internet.

There are four types of users on the Golem Network:

- Providers supply computing power to the network and get paid for doing so.
- Authors build the software on the Application Registry and are paid for their software.
- Requestors rent the network's computing power and the software built on it and pay for doing so. And
- Validators are the ones who make sure that malicious programs do not damage your computer. They analyze & test the software that Authors put on the Application

---

[257] https://filecoin.io/
[258] https://www.ethnews.com/whisper-and-anonymity
[259] http://devp2p.com/

Registry, and create lists for software they trust and for dangerous software.

With such world computer on the horizon, the Ethereum platform gains a name of Ethereum virtual machine (EVM), which refers to the part of the protocol that handles internal state and computation.

Bitcoin, Blockchain & Fintech

Other blockchain applications

Chapter 9 Fintech & the future of financial market

Fintech is short for Financial Technology. It is to apply the blockchain technology, internet, mobile technologies, Big Data, artificial intelligence and other new technologies to the financial services, such as mobile payments, money transfers, loans, fundraising, asset and wealth management, digital currencies and many others. The Fintech using blockchain technology is only a subset of the Fintech.

The drastic rise of investment in Fintech in recent years is due to the advancement of these new technologies, especially the blockchain technology. Fintech creates a proliferation of new applications delivering financial services directly to the end customers.

Throughout history, the advancement of techniques of transactions, from the invention of the money to the invention of paper money, always triggered the revolution in the financial industry. Money and finance are essentially data flow of the economic activities. The financial industry cannot exist without the record keeping, processing and transaction of the financial data.

The advancement of data flow greatly reduces the operating costs, improved service of the financial industry. The invention of the computer and modern communication technology pushed financial technology to a new high in term of speed, cost, and efficiency. The arrival of blockchain technology and other above-mentioned technologies will reshape the traditional business model and continue to bring cost reduction, speed, security of financial transactions, asset transfer and business opportunities.[260]

The traditional financial industry already saw a rapid change in the last five decades. Today, the change is accelerating. The new Fintech companies are bringing more improvements in service and better meeting customer demands. As the technology of smartphone becomes omnipresent, the birth of the blockchain technology spurs additional momentum to the development of the financial technology.

In 2015, the worldwide Fintech investment exceeded US$20 billion for the first time. It still sounds small, but it is growing fast. There are over 12,000 Fintech startups today with an average valuation of $4.4 million. It is a phenomenon rival the early days of the internet.

Fidelity National Information Service (FIS), one of the world's largest providers of financial technology solutions,[261] published a report. It shows that 41% of the companies participating in the survey are testing or implementing blockchain technology to increase revenue stream, 47% are using blockchain technology for collateral management, 42% use blockchains for regulatory reporting, and 36% have implemented blockchain technology for clearing and settlements.

As the word "Fintech" implies, its development has participation from both financial companies and technology companies. IBM, Google, Intel, eBay are the big names investing in the Fintech. Manufacturing companies are also interested in Fintech because it can streamline manufacturing, financial transaction, and supply chain related activities. Many of the baseline technologies driving the Fintech are also driving the Industrie 4.0 – next generation of manufacturing. Many governments embarked Fintech projects for such applications as

---

[260] "Business opportunities for Swiss Fintech companies in Russia", Switzerland Global Enterprise, https://www.s-ge.com/sites/default/files/cserver/mig/sites/default/files/censhare_files/paper-fintech-opportunities-russia_3.pdf
[261] https://www.fisglobal.com/

Other blockchain applications

digital currency, land title registration, etc. It is truly a global effort from multiple sectors of the economy.

Fintech embraces all aspects of the financial systems, including payment and transfer, lending and financing, retail banking, financial management, insurance, credit rating, market, and exchanges. As a result, Fintech is shaking up all sectors of the economy in a big way. Its disruptive technologies are challenging the traditional business model by offering everything from peer-to-peer payments and loans to investment management and crowdfunding.

While Fintech start-ups and technology providers wield huge influence, they currently lack the scale, experience, established client trust and regulatory-backing to embark the financial services on their own. At the same time, the financial institutions want to maintain and improve their position by keeping customers happy, increasing revenues, and reducing costs, becoming more efficient, and maintaining a competitive, well-run business.

The financial industry started to take notice Bitcoin in the early years of Bitcoin around 2010 but did not show much interested in it. It looked upon Bitcoin as an investment instrument. However, when the underlying technology such as blockchain and MDL's were widely discussed and other applications besides the cryptocurrency, such as the Ethereum platform, was developed, the financial industries took the notice.

The financial services industry has the highest ratio of IT spending versus revenue, US$200 billion in 2015, most of which is spent in maintenance rather than creating new services. Blockchain technology provides an opportunity to address this issue. Banks are exploring the opportunities with these Fintech start-ups through capital investment and joint development projects. By doing so, banks ensure that they are positioned to tap into Fintech potential, and leverage the technology expertise of non-bank Fintech players. They recognize that failure to do so bring the risk of being obsolete.

Innovation is emerging in many different business areas within the finance industry, such as retail, market players, lending, payments, messaging, security and foreign exchange. The changes are expected in new and improved solutions, data management, security, and modular IT. The Fintech platforms in development are on both P2P and Cloud-based. Cloud-based solutions are flexible, cost-effective and can scale up to accommodate growing demands, enabling businesses to build and adapt their operations more effectively and efficiently.

Application programming interfaces (APIs) enable the interaction between the user and the service as well as two or more online-connected services, providing the opportunity to build solutions that integrate and combine different services and data sources. Both banks and non-banks have demonstrated the high degree of adoption. New cloud and API technology have been instrumental in enabling the Fintech start-ups to disrupt established players and accelerate change.

Big Data analysis can detect fraud based on customers' spending pattern and geographic location. Artificial intelligence has made it possible to effectively analyze and interpret vast, complex sets of data; uncovering untapped patterns and trends.[262] It allows banks create solutions that are more effective, optimizes business processes and provides value-added services to meet client needs, and

The generation, storage and transfer of a large amount of financial data online prompt more security concerns. Biometric security techniques find wider adaption in smartphones and other mobile devices. Blockchain technology brings additional security to the system.

As Fintech companies continue to introduce new digital capabilities, the banking sector is facing a paradigm shift. The banks that fail to keep up will fade away. Indeed, the financial service industry sees the potential for change in the banking industry as positive and reassesses their strategies and business models. A partnership was born out of the blockchain technology companies and the financial institutions.

---

[262] "How AI is transforming the future of Fintech", Rich Wordsworth, http://www.wired.co.uk/article/how-ai-is-transforming-the-future-of-fintech

Other blockchain applications

Section 9.01 Financial vs. technology companies

Blockchain technology was hyped as a solution for everything. The financial institutions are both afraid of the negative impact it might have on them, and trying to understand how they can fit in, or better how they can benefit from it. The lure of blockchain is in its method of securely verifying and tracking transactions cheaply and automatically.

The threat of blockchain technology, from financial institutions' perspective, comes from much-hyped blockchain's ability to bypass a trusted third party to perform transactions. Blockchain enables non-bank entities to enter markets traditionally that are the domain of banks and other financial institutions.

It turns out that Bitcoin is only one of the many applications of the blockchain technology. The applications built on blockchain do not have to resemble Bitcoin in any meaningful way. For an analogy, the internet is built on the protocols called TCP/IP; however, there is a wide variety of applications built on top of TCP/IP. Many of them bear little resemblance.

Financial institutions found out that blockchain technology has the characteristics that can greatly enhance their hold on the financial services. The key is the Permissioned blockchain. With the Permissioned blockchain, the financial institutions can build applications that solely to meet their own requirements and retain the ownership. This greatly increased their interest level.

Fintech service providers working with the financial institutions aim to capture customers by offering a more intuitive way to access products and services. As the smartphone and internet penetration continues to increase, the financial service is primed for technological adoption and breakthroughs in the workplace.

In 2017, the Fintech industry is still in the early stage of development. Companies are trying to identify opportunities for innovation. In the next stage, we will see more development of the application platform and the proof of its concept, feasibility, and impact on existing systems, such as the supply chain system. Some will work as intended and others will not.

Understanding the regulatory and data security are also important. As of 2017, the regulations do not cover cryptocurrencies. A case in the point, the DAO crowdfunding, and its subsequent hacking caused some anxiety in the investors and raised many questions in its regulatory issues. It should be noted here that DAO hacking occurred not because of blockchain's weakness, but because of its implementation, which has been discussed in Section 4.06.

Because the blockchain eliminates errors and duplication, it is ideal for transforming a host of digital processes. Decentralization of trust has introduced possibilities to make processes such as cross-border payments, trading, and settlement faster, more reliable and less costly. Data integrity ensured by chronological storing of data enforced with cryptography reduces the risk. This, in turn, reduces the compliance burden and cuts regulatory costs with the KYC initiatives.

Even before the appearance of the blockchain, banks already saw an increase in competition from nonbanking entities, especially high tech and e-commerce players in areas such as mobile payments and lending. The blockchain is likely to intensify such competition, as it will reduce technological barriers for digitally perceptive nonbanking entrants.

A blockchain-based system can allow companies to become market makers and open up cash in exchange for completing a cross-border transaction at a lower rate. Blockchain applications also open up alternative funding methods beyond the traditional channels controlled by the venture capitalist, private equity, or IPO. Likewise, blockchain applications also open up other lending channels.

Banks, while they cannot prevent other people from chipping away their business, will have to find other innovative and efficient ways to compete. They are also creating their own versions of blockchain applications and could make quick inroads into their traditional strongholds. The blockchain will create a new set of opportunities for banks to collaborate with startups exploring niche business areas, which integrate Internet of Things (IoT) and/ or smart devices to carry out autonomous transactions through smart contracts.

Other blockchain applications

Section 9.02 Global Fintech landscape

Hot money is pouring into Fintech. Venture capital funding for Bitcoin and blockchain startups reached $2.5 billion in 2016. Meanwhile, many top U.S. and European banks are exploring blockchain applications by either collaborating with startups or creating innovation labs to test their proofs of concept.

Areas of focus for banks and startups include cross-border payments, trading activities, custody services, and customer behavior analysis. Santander, for example, claims to have identified 20 to 25 use cases, with a focus on international payments and smart contracts. Barclays is reportedly focusing on 45 internal use case experiments, [263] while Citibank has created its own version of Bitcoin, called Citicoin. [264] Startups focusing on non-financial use cases have seen a jump in numbers, with new entities entering the market each year. In fact, the number shows that nonfinancial use cases outnumber financial ones, indicating that the applications are not confined to the financial one, and many applications allow real-world assets be linked and traded in the blockchain.

Blockchain's disruptive nature comes from its ability to transform almost any process, from the basic documentation to settling complex contracts across geographies. This inherent capability is alluring to finance and banking decision makers, who believe its disruptive power is good for their industry.

Blockchain's transformative effect will extend to the global financial system. Currently, security trades take two to three days for payments and securities settlement. Employing blockchain technology with a decentralized ledger can greatly speed up the settlement process. It will have a transformative effect on the capital markets.

---

[263] "How Barkley stole the blockchain spot light in 2016", Bailey Reutzel, https://www.coindesk.com/barclays-stole-blockchain-spotlight-2016/

[264] "Citibank is working on its own digital currency- Citicoin", John Biggs, https://techcrunch.com/2015/07/07/citibank-is-working-on-its-own-digital-currency-citicoin/

This benefit extends to very complex instruments, such as derivatives. There are many incentives for banks and financial institutions to deploy blockchain technology in capital markets, such as lower operational cost, faster trade settlement. Executing international trades will become as easy as domestic trades. Decentralizing the clearing process will eliminate the considerable amount of trading risk. Trust level will increase with all transactions recorded transparently on a distributed ledger. The real-time transaction would eliminate counterparty risk. Regulatory reporting becomes easier with easier access to transaction information for regulators.

A blockchain startup R3 attracted 42 international banks and financial institutions. [265] R3 enlisted 11 bank partners to develop a peer-to-peer distributed ledger. It has established industry standards and protocols for blockchain in banking. It also develops commercial applications for banks and financial institutions. R3's effort to create industry standards is a small but significant step toward creating interoperability of blockchain solutions across the financial system.

R3, which opened the first tranche of US$200 million financing exclusively to members of the consortium, symbolized a new trend for the mainstreaming of blockchain technology. The company already counts its notable clients, such as the government of Singapore, the Bank of Canada and other national financial institutions. R3's proprietary ledger can develop applications and support the infrastructure network for financial services firms and technology companies to build their own ledger-based applications and services. Besides R3, other projects such as the Hyperledger Project, Post Trade Distributed Ledger (PTDL) and Digital Asset Holding, etc. are doing pilot runs for blockchain prototypes.

Due to the accelerated speed of globalization in the last two decades, especially after China joined WTO, the world's existing financial system is getting more strained and adding a cost to the trade and service. For example, from 2006 to 2016, the world export of the manufactured goods grew from US$8 trillion to US$11 trillion, while both the world export of agriculture

---

[265] https://www.r3.com/

Other blockchain applications

products and travel and other commercial services have increased 70% during the same period.[266]

The emergence of MDL technology comes at a time when the financial services industry is in need of further improvement. Innovations in digital technologies, MDL technology, and smart contracts have the potential to enhance capital markets infrastructure significantly.

Section 9.03 Technology-driven Fintech

Fintech arises due to the arriving of several disruptive technologies in the first decade of 21st century. Blockchain technology is one of them, probably the most important one. Nevertheless, other technologies also have a profound impact on the development of Fintech, such as mobile technology, Big Data, and artificial intelligence.

Since the invention of iPhone in 2007, mobile technology is more than just the cell phone communication. It is an internet on the go. It is a portal of the internet in one's palm. Tens of thousands of apps are available to harness the power of mobile internet. Many of them are driving the Fintech.

The technology-driven Fintech revolution is evident in two prominent examples: Alipay and WeChatPay. In 2016 alone, these two apps handled US$5 trillion of online payment in China, rivaling US$8.2 trillion payment handled by VISA worldwide. Other business, such as bike sharing, sprang to live because of the mobile payments. One can rent a bike as easy as scanning the QR code on the bike using mobile payment app to unlock it. Two of the largest Chinese-based start-ups, Ofo and Mobike, have a combined more than 13 million bikes and have each raised at least

---

[266] "Trend in world trade", WTO, https://www.wto.org/english/res_e/statis_e/wts2017_e/WTO_Chapter_02_e.pdf

US$1 billion.[267] The Alipay app offers the investment opportunity in online money market fund of US$217 billion, the largest in the world. Such is the power of mobile technology in the financial market.

Likewise, e-Commerce rising out of the internet also transforms the retail business. E-Commerce giant, Amazon, has a market cap of US$570 billion almost twice as big as Walmart (US$290 billion). Chinese e-Commerce leader, Alibaba, also has market cap US$440 billion, not far behind that of Amazon. With the arrival of the blockchain, more changes are coming.

Using mobile phone to rent a bike is an example how technology is transforming the way to initiate and process transactions. With the arrival of blockchain technology, all forms of the payment will change, including the newly arrived mobile payment.

Fintech startups are playing a significant role in altering the financial service market. This presents both a challenge and an opportunity for banks. Fintech start-ups and e-Commerce giants have made a quick inroad into the payment market bypassing banks. These new players are taking payments to the next level of performance in transaction speed and convenience. The biggest impact is on the retail payments.

An increasing number of new non-bank financial service providers are exploring foreign exchange and remittances market, a huge market dominated by the traditional banks. The blockchain platform enables to exchange money in real-time, reducing the exchange rate risk. There are already foreign exchange service providers entered the market, offering minimal costs and P2P business models such as WeSwap.[268] It bypasses banking

---

[267] "Cash is already pretty much dead in China as the country lives the future with mobile pay" Evelyn Cheng, CNBC, https://www.cnbc.com/2017/10/08/china-is-living-the-future-of-mobile-pay-right-now.html

[268] "Innovation in payment", BNY Mellon, https://bravenewcoin.com/assets/Industry-Reports-2015/BNY-Melon-innovation-in-payments-the-future-is-fintech.pdf

Other blockchain applications

networks and is more efficient. Meanwhile, the real overhaul of the foreign exchange market must come from governments. In Europe, initiatives such as SEPA (Single Euro Payments Area) and TARGET2 (Trans-European Automated Real-time Gross Settlement Express Transfer System) intend to make cross-country transfer as easy as a domestic transfer does.

Fintech startups are playing a significant role in altering the financial service market. However, the founders of Fintech startups are usually technology-savvy entrepreneurs, who do not have the expertise or the exposure in the financial service market. Fintech startups can bring in-depth technology and expertise, but lack banks' knowledge of the intricacies and practicalities of functioning payment systems and whether new concepts could be realistically applied in the real world. Banks will be able to offer guidance regarding regulatory requirements and security standards – an area that Fintech startups have little exposure. Banks also have access to large pool of clients. In this background, banks and Fintech startup find out that they need each other.

Fintech Innovation Lab, administered by Accenture, and sponsored by the Partnership Fund for New York City, Credit Suisse, Goldman Sachs, J. P. Morgan and Lloyds Banking Group, and many others, is a program designed to help startup companies to refine and test their value proposition. It also has partnership arrangements with many startup Fintech companies.

Barclays encourages Fintech startups with a program called Accelerator Programme, which provides Fintech start-ups with funding, office space and access to Barclays' APIs and data. California's startup accelerator company, Plug and Play, connects VCs and angel investor to the startup community. It counts Citi Ventures, Citi's venture capital subsidiary, as the financial supporter of the "The Plug and Play Fintech Programme". [269] London based Level39 is another example of the accelerator of technology to Europe's most high-profile Fintech firms.[270]

With the Permissioned blockchain technology, Fintech will supplement the current banking system rather than replace it. For example, the bank account is still indispensable in order to use

---

[269] http://www.citigroup.com/citi/news/2014/141202c.htm
[270] https://www.level39.co/news/welcome-level39-2/

most Fintech services. Banks must now priorities how to adopt a new technology-focused strategy. Despite the fact that non-bank payment service providers are eroding bank's traditional business, few of them want to take on the heavily regulated financial service on its own. Most of them seek partnership with the existing financial institutions to develop new Fintech applications.

Adopting a Fintech-friendly strategy can position banks at the driving seat of the Fintech revolution, ultimately enabling them to provide relevant, user-friendly solutions that present real value and meet the evolving needs of their clients. Banks have always invested in the development of more sophisticated technology capabilities in order to improve payment operations and client service, reduce risk, lower costs and establish a competitive advantage. Such a tradition will serve them well in the future.

Many established banks are investing heavily in Fintech and focus on exploring the potential they have to offer the global payments arena. They developed strong relationships with Fintech players and established a Fintech strategy to re-engineer the process of payments and other assets, such as security, bonds, loans etc, using blockchain and Big Data technology. They also identify key value-adding elements in financial services (such as risk, costs, transparency, and speed) for improvement.

Such a development for the banks requires substantial effort from the proof of concept to the deployment on a larger scale. Banks have to choose from wide varieties of technology applications that may or may not serve their interest. The process of choosing applications can be daunting because the technologies are evolving and their standards are migrating. Worst of all, the new regulations are not defined.

Traditionally, the financial services industry has one of the highest ratios of IT spend as a proportion of revenue. In 2017 alone, the IT spending of the banks in North American, Europe, and the Asia Pacific amounts US$215 billion,[271] mostly on system maintenance rather than on the development of new services. The

---

[271] https://www.statista.com/statistics/554889/it-expenses-of-banks-by-region/

Other blockchain applications

banks are hopeful that the deployment of Fintech may provide some saving in IT.

The new technology is also becoming a great equalizer. Emerging market will leapfrog to adopt the newer generation of technology. The impact demonstrates quickly in the retail payments sector, as it is evident in China. Banks must also adapt quickly to the mobile payment trend.

Not only emerging market benefits from the technology equalization, Small and Medium Enterprise (SME), traditionally underserved by banks, can also benefit from the technology equalization. The emergence and growing popularity of non-traditional forms of finance such as Supply Chain Financing (SCF) and P2P lending can channel much-needed financing to this underserved sector. A growing number of Fintech start-ups are also enabling SMEs to access payment services that were previously unavailable to them. The development of new retail payment system exerts enormous influence over the future path of corporate payments. The traditional players and Fintech pioneers establish collaborative partnerships to leverage the best elements of both parties, and thereby deliver optimal solutions.

Many countries have introduced real-time payments solutions. In Europe, MyBank is a payment authorization solution that enables users to authorize payments via the online banking portal of their own bank across Europe.

In the UK, a payment system called Paym enables customers of many UK banks to make payments directly to each other's accounts. Currently, the use of mobile payment is much smaller in the corporate and wholesale payments sector. However, the volume will increase in view of its many benefits.

Chinese mobile payment market has already exploded to US$5.5 trillion in 2016, almost 50% of China's GDP and the same size as the US credit/ debit card market.

Consumers in retail banking are also benefitting from the development of payment systems that run in real-time rather than via the traditional method of batched processing. This, in turn, has fuelled further innovation, enabling consumers to conduct payments without the need for credit or bank cards; instead of using service layers that run on top of existing real-time payment infrastructures (e.g., the UK's Zapp, which runs on top of the Faster Payments service).

The trend to real-time payments poses serious technical challenges to the global banking infrastructure, specifical linkages to anti-money laundering (AML) and reporting databases, as well as customer accounts payable/receivable and reconciliation. Without the blockchain technology, it is almost impossible to conduct huge volume of global transactions in real time without running into the problems of money laundering.

One of the hurdles in the implementation of the new payment system is the lack of a standard. The establishment of standards is an important enabler in the growth of national and international payments infrastructures. Common standards must exist so that new services can extend the existing systems without glitches. ISO 20022 (the universal financial industry message scheme) [272] and XML standards [273] (electronic data exchange standards for exchange of transactional information between trading partners) are finding wider adaption.

With so much effort going into the payment sector of the Fintech, a new landscape may emerge dramatically in the coming decade.

There is also a new initiative called Bank Payment Hubs (BPHs), [274] which brings together different elements of bank payment systems to better manage payment flows and improve flexibility. It also allows AML to integrate across banks and enable banks to better handle the demands of faster payments and manage risks. In addition, with BPH, there is an improved transparency and availability of data, which can add value to both banks and end-client businesses in terms of process management, cash management, and cost savings. BPHs also move bank payment systems to industry standard messaging, enabling them to connect more seamlessly with outside channels and partners.

---

[272] https://www.iso20022.org/
[273] https://www.gs1.org/edi-xml/xml-advanced-remittance-notification/3-0
[274] "What is payment hub", Lisa Perales, http://blog.powertopay.com/what-is-a-payment-hub

Other blockchain applications

Fiserv,[275] a data processing Fortune 500 company based in Wisconsin, US, focuses on the financial services industry. It has been providing the tools such as billing and payment, ATM processing, auto loan solutions, risk management using Prologue Financial general ledger software to serve financial institutions. There is a growing number of new banking platform providers emerging, such as Bank of America Merrill Lynch (BAML)[276] and Aveloq,[277] offering ready-made platforms capable of plugging into new systems, such as SEPA and Faster Payments. These ready-made platforms are application friendly and extremely easy to implement. They substantially increase the speed to deploy the new technology. Many banks are adopting such platforms, rather than building new platforms from scratch on their own.

Section 9.04　Build MDL for financial services

There are a hundred types of financial services. All serve different purposes and have different complexities. Possible applications of blockchain in financial services not discussed previously include private banking, capital market service, certificates of deposit, bonds, derivatives, voting rights associated with financial instruments, commodities, stocks, credit data, mortgage or loan, P2P lending, donations, airline miles, business license, business ownership/incorporation/dissolution, chain of custody, etc.

In general, the trusted third parties in finance services provide three functions: validation, safeguarding, and preservation. In principle, any financial services, which perform these three functions, can deploy the blockchain.

In developing a blockchain application, the important questions to ask are how to construct the blockchain that will best meet the application requirements? What kind of consensus mechanism is the best? Is it Proof-of-stake, proof-of-work, Proof-of-Concept, dBFT or hybrid? What elements of the trusted third

---

[275] https://www.fiserv.com/index.aspx
[276] https://www.bofaml.com/content/boaml/en_us/home.html
[277] https://www.avaloq.com/

party are to be replaced by blockchain? How are transactions authorized? True peer-to-peer or merely decentralized? Are all nodes equal and performing the same tasks, or do some nodes have more power and additional tasks? Are tokens needed? Is the blockchain application in question going to interface with other blockchains? Does it need sidechain? Is there a need to separate Identity MDL, Transaction MDL and Content MDL? What technical choices are on cryptography standards, peer-to-peer arrangements, guaranteed distribution approaches, partial cryptography, programming languages, communication protocols, etc. Is it a private or public? Is it Permissioned or Permissionless? Is it P2P based or cloud based? How scalable it needs to be?

For example, if the application already has trusted third parties, such as banks, or regulators, such as government, PoW will not be required. If the scalability is needed, PoW cannot be used. Furthermore, if the trusted third party is still present, does the blockchain still offer any advantages? Only when these questions have definite answers, the framework design of the blockchain can be defined.

After defining the scope of MDL in question, the next task is to build the blockchain itself and a small suite of software providing an interface to MDLs for tasks such as selection and storage of documents, document encryption, sharing keys, viewing the MDL transactions and viewing the MDL contents subject to encrypted limits. A suite of software is needed to test various options for MDLs. It should permit testing a variety of MDL configurations and simulated situations. The outputs are to be shared and revised with participants.

The InterChainZ project was a consortium research project, pioneered by London based Z/Yen to provide a real learning experience on MDLs. The project intended to demonstrate how MDLs could work for financial services. The tool used is the online demonstrator of InterChainZ, an online MDL, which can be configured and explore how they might work in a different environment of use. The outputs are interlinked

Other blockchain applications

MDLs along with software, explanatory materials, and website information.[278]

MDLs incorporate trusted third parties for some functions can have significant potential in financial services, such as know-your-customer (KYC), antimony-laundering (AML), insurance, credit and wholesale financial services. In addition, the encryption of MDLs produces immutable records.

InterChainZ demonstrated many potential applications, for example, data ledger, identity application, insurance policy placement, large-scale archive and voting validation. It also works with partners on specific case uses. For example:

- Validation: It worked with an accounting firm to use MDL as identity validator. A third party, e.g., a bank, confirms the validation of the identity and financial information. Such a validation service is useful to individuals who need to comply with AML or KYC requirements.
- Audit: It worked with corporate due diligence officer for corporate credit audit. In this demonstration, MDL functionality allows companies to validate their identity and finance report. A trusted third party reviews the information, confirms and adds it to MDL. The potential creditors or business partners are provided with a public key so that they can confirm the validation and view the information.
- Insurance policy placement: The MDL for this application contains validated insurance history and relevant data of an individual or business. When he or it applies a new insurance policy, he or it shares the key with the insurance company. Since MDL is a trusted database, it saves the insurance company to verify historical data relevant to the applicant. New policy details are added to the MDL as an update.

Z/Yen also explored InterChainZ storage options and network architectures using both Content MDL and Identity MDL and tested the scalability of InterChainZ. InterChainZ MDL has

---

[278] http://www.zyen.com

the advantages of the centrally controlled ledger, simple approval rules, fast entry to the ledger, simple implementation and maintenance, reliant on the single trusted third party, and not dependent on specific nodes to be available. With the burden of PoW, InterChainZ can validate 3,000 to 5,000 transactions per second not as fast as credit cards but close enough.

Section 9.05 Blockchain technology for banks

Financial institutions need to identify opportunities for innovation using blockchain. Financial service operations are vast and complex. To identify areas of traditional operations suitable to apply blockchain technology with clear benefits without negative impact requires serious investigation, because many implications of such change are not well understood.

Since blockchain's operation is essentially transaction and ledger, it seems that ledger and transaction related operations are natural candidates for the blockchain application. In the banking operation related to the shipping business, documentation tracking, bills of lading, letters of credit, load line exemptions, etc. are transactions and ledgers that can use MDL technology.

In a supply chain related operation, payments form a Chain of Custody (CoC), a chronological paper trail of transaction flow. It is what MDL does the best. By using Identity MDL and Transaction MDL, banks can immediately reap the benefit of the improved security of the stored identity, improve portability of data and reduce the time taken for KYC efforts. The data in MDL are much more secure than in the traditional database.

The benefits of MDL applications are more than just data security. Since the bank has non-stop operations, any glitches caused by database going down can be costly. The distributed nature of MDL ensures that the database never goes down.

However, for the banks to make the transition from traditional system to MDL based system is not easy. The most challenging task is not the construction of MDL. The real challenge is to manage the transition from traditional system to MDL and to test out against the real-world environment. Banks

Other blockchain applications

need to create plans to enable blockchain technology to co-exist with their traditional systems. Only when blockchain is mature and robust enough, it can replace existing banking systems. It also needs a common protocol that enables interoperability.

Banks planning to move their processes to blockchain need to start by assessing how interoperability can advance their blockchain objectives. Since most likely banks themselves do not have the expertise to the blockchain technology, they will employ blockchain technology firms to do the development and implementation. Together, they will define protocols and standards upon which the future of blockchain applications will be built. By measuring the results against expectations, banks will be able to refine the application and use this knowledge for future application development.

Since the financial market is heavily regulated, regulations will be an important consideration in the MDL implementation. It is essential to grasp the implication of the regulatory environment before the MDL development. For example, the current regulatory framework has no provisions for accommodating a technology that will eliminate intermediaries. Storing financial data on the distributed network in different countries will also face compliance issue. There is no legal framework for smart contracts today. How to resolve the dispute arising from smart contracts requires new legal definitions. So are the validation, safeguarding and preservation functions not performed by the third party but by blockchain platform.

If the legal framework of MDL operation is established, financial institute operations can be greatly simplified by replacing these functions by blockchain. By doing so, it can lower the barriers and costs of setting up trusted third-party services, and perhaps lead to increased demand.

Identity verification, authentication and data management could streamline many traditional operations of the financial institutions. For insurance companies, the streamlining of digital authentication and better management of personal data and history disclosure could translate into more direct and efficient relationships between insurance companies and their customers. Over time, this could bring additional benefits by reducing identity and claim frauds.

In KYC and AML processes, an identity MDL could transform service levels. Global leaders including UBS, Deutsche Banks, JP Morgan and Bank of America Merrill Lynch are testing blockchain applications to improve workflows and reduce costs.

Trade financing will become easier when it comes to applying blockchain technology. A trade financing workflow based on MDL would allow both the carriers to issue a bill of lading and banks to issue a letter of credit as digital assets on the same blockchain.

Decentralizing document verification would allow companies to execute the most updated version of documents and verify their authenticity. The verified documents can be shared with third party requestors.

Numerous potential applications are discussed in Chapter 6. Due to the limited scope of this book, we cannot cover all the possible applications, and new applications are being uncovered as the technology development continues.

Section 9.06   Mobile wallet

Mobile wallet or mobile electronic wallet is a new form of payment using mobile devices, i.e. mobile phones. Such payment system is prevalent in China but has not been as popular elsewhere. However, such a trend is evident and sooner or later, it will replace credit cards and cash in the near future, as it has already happened in China.

The mobile wallet is an app in the mobile devices to store the information of payment method, which can be credit card or link to the bank account. In China, most of the mobile wallet uses third-party payment system. In the US, some popular mobile wallets are Apple Pay, Google Pay, and Samsung Pay. Many financial institutions, such as Chase, Capital One, PayPal, Wells Fargo Bank also have mobile payment apps. Even some retail stores, like Walmart, Starbucks, offer such a service.

Likewise, Europe also has developed mobile wallet system rapidly. There are many competing wallets operated by various types of domestic and international entities in different countries. An organization called Mobey Forum Digital Wallet Working

Other blockchain applications

Group has created profiles of 49 prominent mobile wallets by their strategy, features, functionality, and technology.[279] Banks operate 26 of them. These include iDEAL of Netherlands, MobilePay of Denmark, BKM Express of Turkey, Swish of Sweden, Vipps of Norway. The other 23 mobile wallets are non-bank operated. While bank operated wallets are mostly serving the domestic market, the non-bank wallets are more successful across Europe. They include Neteller, Skrill, PayPal, SEQR, Masterpass. Yoyo and Amazon Pay. 27 of these mobile wallets are credit card based, while the remaining is Automated Clearing House (ACH) network based, which transfers funds between banks.

The advantages of the mobile wallet are ease of use, without the need of carrying multiple credit cards. Many of the sponsors of mobile wallets also provide rewards as incentives to the customers. Some mobile wallets also provide P2P solution. At merchant POS, some mobile wallets use Near Field Communication (NFC) solutions. Others use QR code, Bluetooth or even Barcode.

One of the major concerns of the mobile wallet is safety. It is inherently less safe than credit cards because the payments go through mobile devices and mobile network rather than dedicated credit card point of sales device. However, blockchain technology can address such a safety issue.

Section 9.07   Crowdsales

Initial Coin Offering (ICO), also known as Crowdsales, is a special form of crowdfunding. It is a new way for the digital currency startups to issue new coins to fund their venture. Typically, the startups release a percentage of the cryptocurrency to early backers of the project in exchange for funding. Because ICO is such a new phenomenon, there is no regulation on ICO process yet and it is prone to be fraudulent. ICO companies essentially bypass the rigorous and regulated capital- raising process required by venture capitalists or banks.

---

[279] https://www.mobeyforum.org/european-digital-wallet-landscape/

ICO offers blockchain startups and projects an easy method of raising capital without the existence of intermediaries and mediators. Some ICO's have raised millions of dollars at very high valuation without a clear vision, user base, revenue stream and even a working product. This would not be possible with traditional angel funding and venture capital firms without an active user base and a solid revenue stream.

It will not be long before the legislation and regulation of ICOs to appear. Like IPO, in an ICO event, the startup firm creates a business plan for its project including the target money to be raised. It tries to sell its business plan to the potential investors. Instead of issuing shares of the startup company to the investors, it issues crypto coins. A successful ICO example is the Ethereum, which launched ICO in 2014 and raised $18 million.

Section 9.08   Wealth management

Big Data, artificial intelligence have already made the transition from concept to application. Artificial intelligence-related products and applications gradually replace the traditional practitioners to provide personalized service to customers. One example is the Robot wealth management, or robot investment advisory.

Robot investment advisory service provides an online wealth management service. It factors in the assessment of individual investors risk tolerance, profit goals and style preferences and other requirements. It uses machine learning generated by intelligent algorithms and portfolio optimization. Test-proven models offer users with the investment advice tailored to the dynamics of the market to provide timely recommendations for investment.

Robot investment advisory charges fewer fees than the traditional investment advisory. The robot is devoid of human emotions, strict implementation of the strategy removed one important uncertainty in the investment. Robot investment advisory also can absorb, process and digest huge quantity of data that human cannot do. More asset management companies now use "robot" advisor to provide services to their customers.

Other blockchain applications

Many start-up companies, such as Wealthfront,[280] Betterment,[281] Personal Capital[282], Motif Investing, FutureAdvisor, Hedgeable, Nutmeg, are providing robot investment advisory services. The demand for such a service is huge.

The intelligent investment startups can command higher valuation in the market. This is the evidence that it is gaining popularity. However, the definition of the industry for the intelligent investment is relatively vague. Some intelligent investment business is merely P2P platform business. Even some truly intelligent investment startups do not have their investment strategy fully tested by the market. The traditional brokerage houses are also offering intelligent investment services. Charles Schwab offers Schwab Intelligent Portfolio service.[283] Based on investor's answers to a short set of questions, Schwab Intelligent Portfolio selects from among the 53 low-cost professionally selected exchange-traded funds (ETFs) to create a diversified portfolio designed to meet investor's specific goal. Merrilledge of the Bank of America also offers similar service.

In Europe, companies like Wealth Horizon, InvestYourWay, and Swanest in the UK, MoneyFarm in Italy, Vaamo and Owlhub in Germany, Moneyvane in Swissland, all offer intelligent investment service.

China also has developed such a market quickly. In China, intelligent investment products belong to three categories: third-party intelligent investment platform, intelligent products offered by the traditional financial institutions, and Internet wealth management applications. These products include A-share investment advisers, wealth management platform, etc. Investment vehicles include overseas ETF, A shares, bonds, funds and so on.

Wealth management is a rapidly growing market. Although it is unrelated to the blockchain, it is an extension of

---

[280] https://www.wealthfront.com/
[281] https://www.betterment.com/
[282] https://www.personalcapital.com/
[283] https://intelligent.schwab.com/

artificial intelligence into the financial market. It is also considered as Fintech.

Section 9.09   A European example - T2S

T2S, short for TARGET2-Securities is a securities settlement platform to centralize settlement in central bank funds across all European securities markets. It is the twin sister of TARTET2 -Eurosystem's cash transfer system. T2S is to integrate highly fragmented securities settlement mechanisms across the borders of many different countries in Europe.[284] By doing so, it can reduce the costs of cross-border securities settlement, which is many times higher than the cost of domestic settlement. It will also increase competition among providers of clearing and settlement services in Europe.

T2S is changing the European post-trade landscape not only by offering an integrated settlement service in central bank money for securities transactions but also because it has brought post-trade harmonization. Harmonization aims to achieve a common set of rules and standards across countries. The combination of TARGET2 and T2S forms the cornerstone of an integrated financial market in Europe Union.

T2S settlement platform integrates the liquidity and balance sheet management, asset servicing and collateral management into a single operating platform. Thus, the implementation of T2S does not end at the cross-border securities settlement, but it is a critical step forwards the final goal of having a unified market for financial services in the European Union.

The Eurosystem launched the project in 2008 and the platform entered operations in 2015. The Eurosystem owns and operates the platform. Before T2S, the Central Securities Depositories (CSDs) and custodians in different countries handled the post-trade activities, such as clearing and settlement of the cross-border security transactions. To make things complicated is that each country has different regulations. They interface directly

---

[284] "T2S", European Central Bank, https://www.ecb.europa.eu/paym/t2s/html/index.en.html

Other blockchain applications

with the market participants. The fragmented system incurs the high cost of cross-border security trade settlement. Therefore, cross-border security trade volume is small.

After implementation, T2S platform unifies the rules and standards and connects all the CSDs across Europe. The T2S platform can also facilitate post-trade activities, thus banks will be able to streamline their operations and produce cost savings.

With T2S, the cost of cross-border settlement cost is lower and it foresees that the cross-border investment volume will increase dramatically. Therefore, the T2S platform will have a positive impact on financial stability and economic growth in EU.

The T2S platform was initiated a year before the appearance of blockchain technology or MDL technology, and developed during the period MDL was still obscure. When T2S is ready for deployment, MDL technology emerged. T2S is forced to look into the new technology and adapt it accordingly. However, at this late stage of the T2S deployment, there are many unknown challenges to the operation of implementing blockchain onto T2S. There are many debates whether to upgrade or amend to incorporate MDL technology now or later and how to restructure the T2S platform using blockchain.

Meanwhile, incorporating still evolving MDL technology into T2S at this moment could be operationally difficult, expensive and risky, because it involves some fundamental changes to the T2S platform developed not long ago. Critics have warned that hasty implementation of an unproven technology may pose risk to the capital market. In addition, the regulatory impediments of blockchain's adoption into T2S are far from clear. The implication of implementing MDL technology in T2S is more than just cross-border trades. The interface between the cross-border MDL based systems with domestic traditional database systems poses a major challenge.

The European Central Bank (ECB) has already invested a lot of money into T2S development and deployment. To incorporate MDL technology, it amounts to redevelop T2S platform over again. It is truly facing a huge dilemma. The success of the T2S project is critical for the further integration in the EU financial markets. To break through the complexities of development, testing and migration have been a daunting task because of too many conflicting interests.

The ECB has considered a range of blockchain models currently under development. These differ in a number of dimensions, including the validation mechanism, network architecture, permissions, the level of data sharing and replication, and cryptographic tools of choice. The possible mass adoption in the market also needs to weight functional, operational, governance and legal aspects. Since the blockchain technology is still evolving, it is not mature enough to meet high requirements in terms of safety and efficiency for such huge application. The safety of financial markets requires better consideration.

Therefore, for the time being, the ECB decided to put off a DLT solution. However, it remains open to considering innovative solutions in the field of blockchain when such technologies are proven and mature enough to be the cornerstone of the Eurosystem financial infrastructure. ECB entered a joint research project with the Bank of Japan to study the possible use of DLT for market infrastructure. The project can help define how new technologies can change the global financial ecosystem of today and ensure that central banks are adequately prepared. In addition, ECB also established a DLT task force. Its objective is to assess the potential impact of DLT on harmonization.

Section 9.10   TechFin

A term not to be confused with Fintech is the TechFin, or simply technology-based financial services. In contrast to the traditional financial services centered on banks and other financial institutions, such as insurance companies, venture capitals or private equity, TechFin refers to a new breed of financial services created by the proliferation and innovation of new technologies. The financial services offered by TechFin are a new type of services, which could not exist without the baseline technology.

Other blockchain applications

The technology-centric companies using the newly developed Fintech dominate Techfin.[285]

Large internet and e-commerce companies have amassed huge customer base and market share that they become a power in the financial market. Most well known companies of this type are Amazon, Netflix, eBay, PayPal, Google, Apple, Hotel.com, Uber, Airbnb in the US and Baidu, Tencent, Alibaba, JD.com, Juzhen Financials, Zhongnan Construction, a property conglomerate, CreditEase, microfinance and wealth advisor, Shanghai Insurance Exchange, Qtum, and many others in China.

TechFin has the characteristics of harnessing Big Data/internet related technology innovation and recently blockchain and artificial intelligence for financial power. Most of the real disruptors, however, are the companies offering end-to-end solutions that cut out existing players. We have seen this in many economic sectors such as Uber vs. taxis, Netflix vs. cable companies, or AirBnB vs. hotels.

Sovereign funds like GIC Singapore or private equity funds like KKR and Blackstone invested heavily into TechFin. TechFin companies receive more funding than Fintech startups.

Alibaba, an e-commerce giant in China, entered the financial services sector with the creation of Ant Financial in 2016. It is estimated that the valuation of Ant Financial valuation in 2017 could be as high as $75 billion, ranking 20th in world largest banks.[286]

The entry of these Fintech firms is a paradigm shift in the financial industry from money centered financial service to data centered financial service. It may be the single most important development in financial services in the recent history.

TechFin creates new financial service markets using the new technologies. However, these new financial services are often

---

[285] "Fintech vs Techfin", Kunal Patel,
https://www.finextra.com/blogposting/14499/fintech-vs-techfin

[286] "Jack Ma's Ant Financial Nearly Doubles Profits Amid Deal Push", Lulu Yilun Chen,
https://www.bloomberg.com/news/articles/2017-06-22/jack-ma-s-ant-financial-nearly-doubles-profits-amid-deal-push

shortcuts to the existing financial services. They embrace not only financial services but also a retail business (such as e-commerce), service business (such as taxi and hotel). For example, Alibaba is not only an e-commerce company but operates many data center worldwide. [287], [288], [289] Amazon AWS (Amazon Web Services) pioneered cloud computing in 2006, and currently has 49 data center worldwide in 18 countries. [290]

The data centers in the early 21st century are akin to the power plants in the early 20th century. The power plant delivers power to each household while data center delivers data to each individual. These data and data centers give TechFin companies a real advantage in the provision of financial services. When financial services become more and more data-centric, traditional financial institutions will have a difficult time to compete with the power amassed by TechFin companies.

These data centers allow TechFin companies to provide far more efficient financial services and improved decision-making. Consumers get the first taste of the targeted ads from Google Ads. With a search engine, Google Ads can deliver ads to the right customers. So will be the financial services in the future. Such targeted and tailored financial services can be a boom for the small and medium enterprises (SME). TechFin companies also open the door of P2P business across national borders, which is the exclusive realm of B2P business.

Section 9.11   Alternative Trading System

---

[287] "Alibaba scouting locations for a second European data center", Misha Savic, https://www.bloomberg.com/news/articles/2017-09-28/alibaba-is-scouting-locations-for-a-second-european-datacenter

[288] "Alibaba cloud is opening its first data center in India", Jon Russell, https://techcrunch.com/2017/12/20/alibaba-cloud-india/

[289] "China's Alibaba opens Silicon Valley data center", Laura Luo, http://www.datacenterdynamics.com/content-tracks/colo-cloud/chinas-alibaba-opens-silicon-valley-data-center/93479.fullarticle

[290] https://aws.amazon.com/compliance/data-center/

Other blockchain applications

Any platform matching buy-and-sell orders but not regulated as an exchange is an Alternative Trading System. It is not new but exists in many different forms: a multilateral trading facility, electronic communication network, cross-network, or call network. While the exchanges trade security, commodity, foreign exchange, and financial derivative products, most ATS's are trading other products. Usually, ATS is registered as broker-dealer and focuses on finding buyers for the sellers and vice versa to conduct transactions. They do not set rules governing the conduct of trading. However, they do play an important role in providing alternative means for liquidity. This is especially true for the securities not qualified for the stock exchange listing, or the unlisted ADRs.

Theoretically, ATS involves willing parties of buyers and sellers. The rules of public trading do not apply. Yet, the boundary between ATS and exchange is blurry. It can happen that institutional investors use an ATS to find counterparties for transactions of the equity, which is on public stock exchanges. Alternatively, a small group of selected investors, e.g. banks, mutual funds, private equity, pension funds, participate the purchase a large block of securities in a private placement by invitation. A private placement may have the intention to conceal trading from public view since ATS transactions do not appear on national exchange order books. Since such a trading involves publicly traded security or assets, it may violate public interest. Cryptocurrency trading is also an example of ATS trading. Therefore, recently regulators have stepped up enforcement actions against alternative trading systems. The SEC introduced Regulation ATS in 1998 to protect investors and resolve any concerns arising from alternative trading systems.

With the advancement of blockchain technology, more and more trading happens outside of exchanges. ATS growth becomes unstoppable. A Delaware-based Delaware Board of Trade Holdings (DBOT) specializing in blockchain trading platform is the first company to seize the opportunity to develop the technology for ATS. DBOT became the first and the only blockchain based ATS fully licensed by the SEC. Seven stars Cloud Group (SSC), a Chinese-based Artificial Intelligence and Blockchain-Powered Fintech Company for the delivery of digital asset securitization solutions, has acquired a controlling ownership stake in DBOT in 2017. [291]

DBOT's technology will be used to power the SSC's blockchain-based NextGen X platform, which focuses on the trading of financial products including ETFs backed by digital assets, that can be tokenized and settled (including digital currency options) via "plug and Play" and blockchain based Initial Exchange Offerings ('IEO") network. [292] Its plug and play nature allows it to be adapted to different ATSs.

IEO is a newly invented term by SSC to differentiate the fact that the offering to be traded is backed by physical and digital assets rather than a promise of a plan. SSC's NextGen X will apply blockchain and artificial intelligence to create a "hybrid solution" for supply chain finance, risk management, and asset-backed digital securitization. By creating "financial superintelligence", it will be able to rate financial risk and price assets before monetizing into issuance and trading through the IEO. SSC plans to establish ATS in the US, UK, Germany, China, Korea, Africa, Singapore, UAE, and Japan and will expand issuance and trading volume via 30 targeted local and regional exchanges and ATS's.

Section 9.12   O2O business

O2O (Online to Offline) is a new and emerging business model, which connects online business to the traditional brick and mortar stores. In the O2O business model, although customers

---

[291] "Seven Stars Cloud Announces 27% Purchase of Delaware Board of Trade Holdings, Inc.",
https://www.prnewswire.com/news-releases/seven-stars-cloud-announces-27-purchase-of-delaware-board-of-trade-holdings-inc-dbot-300573871.html

[292] "Seven Stars Cloud Announces 27% Purchase of Delaware Board of Trade Holdings, Inc. (DBOT)", The NY Times, http://markets.on.nytimes.com/research/stocks/news/press_releas e.asp?docTag=201712200914PR_NEWS_USPRX____NY7465 6&feedID=600&press_symbol=46908

Other blockchain applications

make purchases online, delivery of the product or service takes place at physical locations. Both the online and offline retailers are adopting an O2O strategy to stay competitive. Today, most of the large retailers, such as Walmart, Target, Macy and many others, allow you to shop online. You can even order online at the kiosk in the store if you do not find the product you want in the store. At the same time, the online retailers are buying into brick and mortar stores. For example, Amazon bought Whole Foods in 2017. The line between online and offline retailers is blurring. The O2O business is also known as the Omni-channel market.

In China, Alibaba formed a new partnership with retail conglomerate Bailian, which operates 4,700 stores across China. It acquired a major stake in department store chain Intime Retail Group [293] for US$2.6 billion, as well as a US$4.6 billion investment in electronics retailer Suning.[294] In November 2017, it invested US$2.87 billion for a major stake in China's top hypermart operator, Sun Art Retail Group.[295] In such a move, Alibaba Group obtained 36.16% in Sun Art Retail Group Limited by acquiring shares from Ruentex. Auchan Retail is also increasing its stake in Sun Art. The transaction gave Auchan Retail, Alibaba Group and Ruentex approximately a 36.18%, 36.16%, and 4.67% economic interest in Sun Art, respectively. [296] In doing so, Alibaba has established offline outlets in multiple retail markets, from the food supply, consumer and department store to electronics. JD.com, counting Walmart as its major

---

[293] "Alibaba Takes Big Step Offline With $2.6 Billion Intime Deal", Bloomberg News,
https://www.bloomberg.com/news/articles/2017-01-09/alibaba-others-to-privatize-intime-for-up-to-hk-19-8-billion

[294] "Alibaba to invest $4.6 billion in China electronics retailer Sunning", Gerry Shih, https://www.reuters.com/article/us-alibaba-suning-appliance/alibaba-to-invest-4-6-billion-in-china-electronics-retailer-suning-idUSKCN0QF0VP20150811

[295] https://www.cnbc.com/2017/11/20/alibaba-invests-2-point-9-billion-in-chinas-sun-art-retail-group.html

[296] https://www.businesswire.com/news/home/20171119005069/en/Alibaba-Group-Auchan-Retail-Ruentex-Form-New

shareholder, likewise financed Chinese supermarket chain Yonghui.

In a demonstration of the O2O concept, Alibaba featured an Augmented Reality mobile game that drove foot traffic to locations owned by Suning, Intime, KFC, and Starbucks, as well as malls in Beijing and Shenzhen and Shanghai Disney Resort. The AR games use GPS enabled smartphones to allow users to move around the neighborhood and complete different tasks, which can be shopping or dining. Users receive instant information such as in-store promotions, in-store pickups of items ordered online, and collaboration on consumer data. The interactive shopping experience grabs customers' attention, curiosity, and transforms shopping into gaming.

Rivals in China such as JD.com are also highly active with O2O. JD.com has opened 1,700 physical appliance stores and planned to have 10,000 brick-and-mortar stores by the end of 2017. In early 2018, JD showed off its 7Fresh stores. The stores use an array of new technologies, such as autonomous shopping carts and Big Data analytics, which help to provide a personal store experience for the customers. The autonomous shopping carts guide the customer through the store autonomously, allowing them to shop hands-free and without having to focus on their purchases. A mobile app and digital payment technology will take care of the scanning and payment procedures. 7Fresh targets 1,000 stores across China, within 3 to 5 years. It will compete with Alibaba's technological supermarket, Hema, which already has 25 stores at the end of 2017.

Baidu also invests in the O2O business for food-delivery, ticketing, and entertainment with a US$3 billion investment in group-buying site Nuomi and leading a US$1.2 billion fundraising round in Uber. Baidu plans to invest US$3.2 billion from 2015-2018 into O2O services. Analyzing customers' purchasing habits and building up more robust user profiles allow the search engine to sell the information to advertisers, influence consumers' choices in O2O spending and receive a cut of each transaction conducted online and completed offline.

Other internet giants such as Alibaba and Tencent also see the massive possibility of future revenue streams from O2O businesses. Tencent has been investing internationally,

Other blockchain applications

diversifying into hundreds of companies focused on gaming, mobile money, and artificial intelligence. Such deals include a US$90 million funding round for Indian health-care information provider, Practo, which provides an online search tool for consumers to find healthcare professionals; and leading a group of investors paying US$8.6 billion78 for a majority stake in Finnish game-maker, Supercell, in June 2016, further cementing its role as a global leader in gaming.

Tencent took a billion-dollar stake in a combined US$4.5 billion private equity investment into Chinese car-hailing app Didi Chuxing. This is in addition to alliances with US's Lyft, India's ride service Ola, and Southeast Asia's ride-hailing startup Grab, and the acquisition of Uber's Chinese network. Tencent also has plans to expand into Hong Kong, Taiwan, Macau, Japan, South Korea, Europe, and Russia. Chinese players going overseas meet challenges different from those in their own turf; as a result, they do not always fare well outside their home ground. Business success overseas requires catering to the peculiarities of international markets and their consumers.

The Offline business is fighting back to go Online. However, the Online to Offline business is moving faster than the Offline to Online business. This is because there are technology innovations involved in the online retail business, which the brick and mortar stores cannot pick up fast enough. As a result, brick-and-mortar stores are struggling to compete with e-commerce.

Section 9.13   Fintech and AI

There are many definitions of artificial intelligence. One of those is the system that has the capability of self-learning. Systems that can learn from experience will become more intelligent. In the financial market, the use of neural networks to predict financial indicators has long been a subject of research. The machines based on a neural network try to imitate living brain. Even the lowest life form can explore, probe, test and learn to complete tasks.

The learning requires experience, where the Big Data comes in. The combination of Big Data and self-learning capability produces the artificial intelligence. The financial Big Data contains people's spending habits, their preferences, their lifestyles, their financial capability, their wealth management objectives, the flow of money, investment trend, economic activity tendency and many others.

When it comes to Fintech, the idea is to create smarter AI that helps the management of financial market and operation. Faster decision-making and deeper learning can predict financial behavior. While many of the Big Data and AI applications in business are not strictly financial services, however, the boundary between financial service and other types of business are getting blurry.

AI will help businesses and banks to tailor products and services to customers' requirements more effectively. It will also discern the patterns of economic trend, consumer patterns. There are already many startups developing AI products and services to improve and expand credit offerings, insurance options, personal finance services, regulatory software or even products beyond financial services. For example companies like Affirm[297],[298], a Danish company co-founded by PayPal's ex-CTO, and Zest Finance use AI for credit scoring and lending applications. Both Baidu and JD.com, two e-commerce giants in China, are investing in Zest Finance.[299]

Digit[300] uses AI to help people accumulate savings and Kasisto[301] offers AI chatbot and mobile app help companies to connect customers. Dataminr[302] searches and identifies in 7/24 the high-impact and threatening critical events worldwide and delivers to personnel, facilities, operations, and interests around the world in real time. Alphasense[303] use AI to do market research and

---

[297] https://www.affirm.com/

[298] "Affirm revolutionizing the loan industry through the use of AI", Maria Thing Nielsen, http://www.icdk.us/blog/affirm-revolutionizing-the-loan-industry-through-the-use-of-ai

[299] "Chinese Search Giant Baidu Just Backed This Fintech Company", Leena Rao, http://fortune.com/2016/07/18/baidu-credit-scores/

[300] https://digit.co/

[301] http://kasisto.com/

[302] https://www.dataminr.com/

[303] http://www.alphasense.com/

sentiment analysis. It launched an AI-powered search engine to provide an intelligent search with critical insights, accuracy, and speed. The search results are significantly more customized and accurate than a simple Google search. Search function becomes more like an expert consulting session.

Sentient Technologies [304] provides AI to identify and answer critical questions in new and groundbreaking ways, empowering people and businesses to make better decisions. For example, it developed an AI-led conversion rate optimization (CRO) to help travel companies optimize their websites and drive the highest possible conversions from visitors. The multivariate testing technology allows users to test multiple different website elements across multiple pages at the same time and rapidly define the most effective combination that drives the highest conversions from customers.

More often, when you dial a service phone number, a chatbot answers the phone. It directs you to different parties depending on your answer. According to SITA survey in 2017,[305] a Geneva-based aviation technology firm, 52% of airlines were planning to use AI programs and 68% of airlines intended to adopt AI-driven chatbots in the next three years. AI-powered chatbots are growing in popularity within numerous service industries.

Another company Numerai[306] develops AI algorithm for trading and investment. Numerai is a new type of hedge fund that is built on crowdsourcing knowledge through a massive network of hedge funds—the system collects hundreds of thousands of financial models and individual predictions. With these data, Numerai builds their own financial models that incorporate the algorithms submitted through the crowdsourced community.

As artificial intelligence starts to propagate at asset management and investment banks, AI Fintech will replace the role of traditional analysts. This will result in more efficient,

---

[304] https://www.sentient.ai/
[305] "How travel companies are using AI to acquire, engage, and retain customers", http://travelwirenews.com/how-travel-companies-are-using-ai-to-acquire-engage-and-retain-customers-2-580212/
[306] https://numer.ai/

accurate and better investment research at lower costs. Different deep learning AI applications have their own unique performance characteristics, which can be further impacted by their different deployment platforms.

There is a need to benchmark Artificial intelligence platform since not all AI's are created equal. Some AI's are more intelligent than the others are. Without benchmarking, there is no way to tell its level of intelligence.

The extent of the capabilities learned during deep learning training is the "inference". Think the inference as the AI's IQ. Measuring AI inference (machine's IQ) of neural networks on a device, which can be a cloud, a network, a computer or a phone, allows us to know the level of IQ of the particular AI device.

Baidu, the dominant Chinese search engine company, has invested heavily in the artificial intelligence. In 2016, Baidu Research unveiled the next generation of DeepBench,[307] the open source deep learning benchmark system across different hardware platforms. DeepBench approaches this issue by benchmarking fundamental operations required for inference. The inference measurement in DeepBench uses a previously trained model to make predictions on a new data set. By doing so, it can measure the deep learning performance after a series of training.

DeepBench is the first tool opened to the deep learning AI community to evaluate how different platforms, or the algorithm/ hardware combination, perform when used to train deep neural networks. Since its release, several companies, including Intel, Nvidia, and AMD, have used and contributed to the DeepBench platform. [308]

---

[307] "DeepBench", Sharan Narang, https://svail.github.io/DeepBench/
[308] "Intel® Xeon Phi™ Delivers Competitive Performance For Deep Learning—And Getting Better Fast", Andreas R, https://software.intel.com/en-us/articles/intel-xeon-phi-delivers-competitive-performance-for-deep-learning-and-getting-better-fast

Other blockchain applications

A better understanding of the performance of inference can improve the design of the chips/ hardware where the neural networks reside. Among different performance benchmarks, speed is key to the training of neural networks. With the ability to measure inference, researchers will now have a standard for the performance of their AI hardware.

In addition to measuring inference performance, DeepBench provides new kernels for training from several different deep learning models. It also sets new minimum precision requirements for training.

AI and machine learning can enable banks to become more competitive in other industries like real estate. Fintech sectors, such as the $55 billion property management industry, are using AI and machine learning to increase returns, lower costs and improve the overall owner and tenant experience for the 50 million rental properties and 100 million tenants in the U.S.

Ecosystem players like mortgage bankers, payment systems, insurers, private banking and wealth management are now working together with technology-led property management companies to serve their clients better.

AI can improve the financial market in many ways. One way is to increase security. Due to AI's capability of machine learning and pattern recognition, AI can analyze large volumes of security data and help to detect abnormal patterns of fraudulent behavior, suspicious transactions, and detect potential future attacks, to keep sensitive information safer. AI can also cut processing times by processing data more accurately and quickly, and reduce human error, validate and double-check data. A Silicon Valley startup, Pixmettle,[309] is developing enterprise AI-based tools, which will help to detect frauds proactively and at a very high accuracy.

Chip companies are also rushing into offering chips designed especially for artificial intelligence applications, which require handling huge workload to process machine learning and deep learning. These players include Intel, AMD, Nvidia, IBM and many others.

---

[309] http://pixmettle.com/

Nvidia's GPU using proprietary NVLink technology, so far the most popular AI chip, in combination with IBM's CPU Power9 can crunch the huge amount of data in AI and machine learning such as image and voice recognition and credit card fraud prevention.

Intel's Nervana Neural Network Processor (NNP), built for deep learning, provides the flexibility to support all deep learning primitives efficiently. It is called the Neural Network Process because the parallelism in its architecture allows neural network-like parameters distributed across multiple chips to accommodate larger models to capture more insight from data.

Advanced Micro Devices launched the newest Epyc server and Vega graphics chips for AI tasks. Google, although not a chip company, has also forayed into designing its own chip named TPU V2 (the Tensor Processing Unit)[310] to power its cloud computing business.

Microsoft offers tools, a turnkey solution, to allow developers to create AI apps on its Azure platform. Such a tool can be used in virtue machine images, networking, storage, the blockchain, security and identity, database and others. Many of these have direct application in Fintech. [311] Many companies are forming partnerships to leverage each other's expertise, e.g. Microsoft is working with Huawei, Facebook partnered with Qualcomm.

AI is by no means limited to the big machines. Even mobile devices are going AI. Apple's new iPhone has a neural engine, so does Huawei's Mate 10. Big companies designing mobile chips, such as Qualcomm and ARM are gearing up to supply AI-capable hardware to power mobile devices. AI hardware promise to offer better performance, better battery life, privacy, and security. AI chips can perform more data crunching on the device rather than sending them to the cloud. This means

---

[310] "Google's next-generation AI training system is monstrously fast",Nick Statt, https://www.theverge.com/2017/5/17/15649628/google-tensor-processing-unit-tensorflow-ai-training-system

[311] https://azuremarketplace.microsoft.com/en-us/marketplace/

improved privacy. AI is more than just hardware; it is the combination of hardware and apps. Both Android and Apple iOS have APIs that can tap the power of the neural hardware. These APIs leverage the capability of deep learning frameworks such as Google's TensorFlow[312] or Facebook's Caffe2.[313]

The R&D spending on AI-oriented microprocessors is ramped up dramatically. Some estimate that the AI chip sales will reach US$35 billion in 2021 from US$6 billion in 2016.[314] AI chips power AI just like microprocessors power computing. It will be omnipresent in the near future.

One of the most common benefits and effects AI has in nearly every industry is its automation opportunities. In addition to AI reducing the manual expense management problem, it can automatically generate expenditure and expense reports quickly, efficiently, and without errors.

AI understands approval workflows and allows companies to restructure and automate the expense tracking process. It also helps to prevent reimbursement fraud and guides organizations with their budgeting efforts, thanks to automated reports.

One of the most valuable benefits AI brings to organizations of all kinds is data. The future of Fintech is largely reliant on gathering data and staying ahead of the competition, and AI can make that happen. With AI, one can process a huge volume of data, which will offer game-changing insights. These insights can help in many ways in the complex decision-making processes.

To succeed, larger enterprises have relied heavily upon the algorithms, automation, and analyses achieved with the help of AI. Because this technology is becoming increasingly accessible and

---

[312] https://www.tensorflow.org/
[313] https://caffe2.ai/docs/getting-started.html?platform=windows&configuration=compile
[314] "In AI Technology Race, U.S. Chips May Be Ace-In-The-Hole Vs. China", Reinhardt Klause, https://www.investors.com/news/technology/ai-technology-u-s-chip-stocks-vs-china/

affordable, it helps smaller startups as well as consumers, giving them the tools to compete with larger players within their industry.

AI is playing an important role in empowering both consumers and Fintech companies. AI-powered personal financial applications allow people to balance their budgets based on their specific income and spending patterns. AI acts as a robot advisor to individuals and organizations to cut costs while boosting the bottom line.

Artificial intelligence is not limited to a behind-the-scenes role in the business world. AI-guided chatbots undertake a variety of internal and external communications, such as the self-service customer-facing tools employed by financial institutions. With the stiff competition in the Fintech world today, it's no surprise that companies are adopting new technologies to stay one step ahead of the competition. Artificial Intelligence allows Fintech companies to eliminate human error while boosting productivity and increasing their bottom line.

Bitcoin, Blockchain & Fintech

Other blockchain applications

Chapter 10 Fintech in China

While London, New York, Frankfurt are the world's financial centers for centuries, China has leapfrogged ahead to become the center of global Fintech innovation and adoption. The Chinese Fintech hubs are concentrated in major cities, such as Shanghai, Hangzhou, Beijing, and Shenzhen.

The speed, sophistication of development and the scale of deployment of China's Fintech ecosystem is breathtaking. A complete value chain for the blockchain sector has emerged in China, ranging from hardware manufacturing, platform and security to application services, investment, media and human resources. The number of blockchain technology companies in China exceeded 400 in 2018. These companies explore ways to apply blockchain in a wide array of areas.

Prompted by huge e-commerce market and growing appetites of consumers, China's technology leaders are revolutionizing many aspects of financial services. While the Chinese Fintech companies are not the most innovative, but unlike the Fintech activities in the West, China's Fintech ventures into many industrial scale applications with the huge user base. Fintech's quick start in the past decade has resulted in several niches – online lending, insurance, etc. A handful of Chinese Fintech firms went public in 2017, led by the first online-only insurer ZhongAn, which claims 523 million customers.

Multiple fundamental factors contribute triggering China's Fintech development - the size of unmet banking needs, government regulatory support, mature mobile payment platform, a huge number of mobile and internet users, large e-commerce market, a growing middle class with disposable income and easy access to capital – fertile breeding ground for Fintech.

In 2017, China amounts to 70% of the worldwide Bitcoin mining activities, and the second largest Fintech investment after the US. China has 8 of the 27 world's largest Fintech companies valued at more than US$1 billion. Fintech Innovators, a collaboration between Fintech investment firm H2 Ventures and KPMG Fintech, announced Fintech100, the leading Fintech companies in the world. [315] Chinese Fintech companies took five of the top ten, including the top three spots. The US took three. Germany and UK took one each. The top three Chinese Fintech companies are Ant Financial, ZhongAn and Qudian.

The world's biggest Bitcoin miner, a Beijing-based company called Bitmain, controls nearly 30% of all the Bitcoin mining power worldwide. Roughly, two-thirds of the world processing power devoted to mining Bitcoin resides in China. Thousands of custom-designed special computers fill these warehouse-like mines, humming non-stop to solve Bitcoin puzzles to receive Bitcoins awards. Although the Chinese government has been trying to crack down Bitcoin mining, exchange, and ICO, so far there are little effects on the Bitcoin. There are 65 ICO events in China in the first half of 2017, collecting over US$394 million, but September's ban on ICOs brought that to a halt. On the other hand, the government is actively encouraging Fintech development with favorable policies and investments. It seems that the Chinese policy is self-contradictory. In reality, China is trying to channel the blockchain technology into the financial and industrial applications, away from the cryptocurrency.

Fintech investments in China including Hong Kong surged to US$10.2 billion in 2016,[316] out of which the venture capitals contributed US$7.7 billion, eclipsing North America's US$9.2

---

[315] "The Fintech100 – Announcing the world's leading fintech innovators for 2017", https://home.kpmg.com/xx/en/home/media/press-releases/2017/11/the-fintech-100-announcing-the-worlds-leading-fintech-innovators-for-2017.html

[316] "Accelerating Fintech in China", Joshua Bateman, https://techcrunch.com/2017/10/08/accelerating-fintech-in-china/

Other blockchain applications

billion, out of which US$6.2 billion from VC,[317] followed by a distant third of UK. Chinese investment continues to dominate in Asia-Pacific.

China's interest in the blockchain technology is more than just Bitcoin. Some of China's biggest firms are betting on its technical capability to revolutionize their businesses. They are aggressively looking for solutions to their existing problems in the newly developed but huge financial markets, such as US$5.5 trillion a year third-party payment market[318] or the US$180 billion a year online lending market, which accounts 75% of the worldwide market share in 2016.

China is also the home of four of the ten largest banks in assets in the world per 2017 S&P Global Market Intelligence report.[319] Chinese banks are highly profitable with average ROE around 17% vs. 9% for the US banks and 2% for the European banks. Because of the insufficient social security safety net and the traditional habit, Chinese tend to save more for the rainy days. The Chinese saving rate is at 46%. In 2015, China in whole has US$27 trillion in savings.[320]

China's largest Fintech company – Ant Financial, valued at about US$60 billion, ranking 20th in the world largest banks if it were a bank. In 2016, it raised US$4.5 billion, making it the largest single private placement in any Fintech firm globally.

---

[317] "Global Fintech VC investment soars in 2016, Lawrence Wintermeyer,
https://www.forbes.com/sites/lawrencewintermeyer/2017/02/17/global-fintech-vc-investment-soars-in-2016/#7e2fe2d82630
[318] "Internet trend 2017", Mary Meeker,
http://www.kpcb.com/internet-trends
[319] https://en.wikipedia.org/wiki/List_of_largest_banks
[320] "With $27 trillion in savings, Chinese are set to change the world", Eda Corran et. al.,
http://www.smh.com.au/business/china/with-28-trillion-in-savings-chinese-are-set-to-change-the-world-20150625-ghy4x1.html

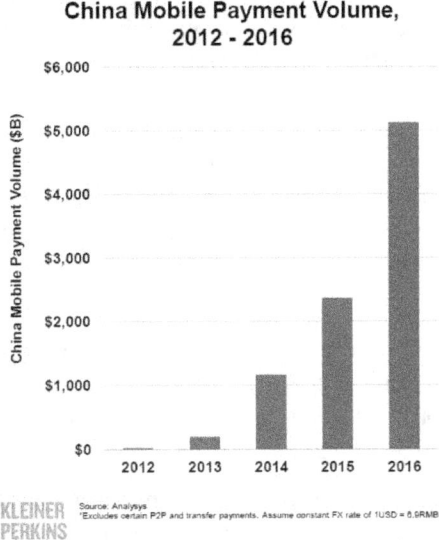

Figure 10-1 China mobile payment market

Two of China's largest internet finance companies, Lufax, a P2P lending firm backed by Ping An Insurance, and JD Finance, a subsidiary of JD.Com, received US$1 billion investment each in 2016.

Chinese government openly supports the Fintech development with policy; funding and government organized consortiums as well as a relaxed regulatory environment. Chinese government allows the new technology and system to evolve first before regulating. While, in the West, existing regulations often prevent the new system to be developed, and new regulations often take years to pass the legislation. The Chinese government also has officially included the scope of digital technology, including internet, cloud computing, Big Data, artificial intelligence, machine learning, the blockchain, into the 13th National Five-Year Plan.

China's Ministry of Industry and Information Technology has organized a special government-industry working group to present a plan for promoting the development and adoption of blockchain technology. The Chinese government also launched the "Internet Plus" initiative in March 2015 to speed up the

Other blockchain applications

integration of internet infrastructure, development of chip technology and high-speed computers, applications of cloud computing and Big Data as part of a wider effort aimed at incorporating internet technology into different sectors of the economy.[321] (Figure 10-2) The plan stipulates development targets and supportive measures for internet and Fintech industries. It aims to hasten the integration of internet/IoT and other derivative technologies, such as Big Data, the blockchain, with different business sectors, including but not limited to traditional financial services.

Chinese government operates more than 750 government-guided funds nationwide. In 2015, it appropriated a massive US$231 billion to fund start-ups, of which US$6.5 billion went to promote digitalization and smart technologies.

China also has a protected financial market as wells as its relatively underdeveloped banking system. On average, there are 8 bank branches per 100,000 people in China vs. 28 bank branches in the US and Europe. Also, the Chinese credit/debit card market is relatively untapped. Interestingly, the Chinese credit/debit card market is about the same size as the US electronic payment market, while the Chinese electronic payment market is about the same size as the US credit card market.

Internet technology has been an important growth engine of Chinese consumer market. Such technology created great companies like Alibaba, Tencent etc. China sees all technologies derived from internet technology as the second opportunity to reap its benefit. A major effort is devoted to the development of artificial intelligence, encryption, decentralization scheme, Big Data, and finance. One of the important infrastructures of Fintech is data center. Data center for data flow in the early 21st century has the same significance as a power plant to the electricity flow in the early 20th century. China is investing heavily in data centers.

Understanding of Fintech in China may provide an insight into China's grand strategy to boost its economy by deploying the

---

[321] "How is Internet Plus altering China", http://www.china.org.cn/business/2016-07/11/content_38851412.htm

latest technologies in the financial industry. Moreover, how such adoption will revolutionize everyday life of its citizens.

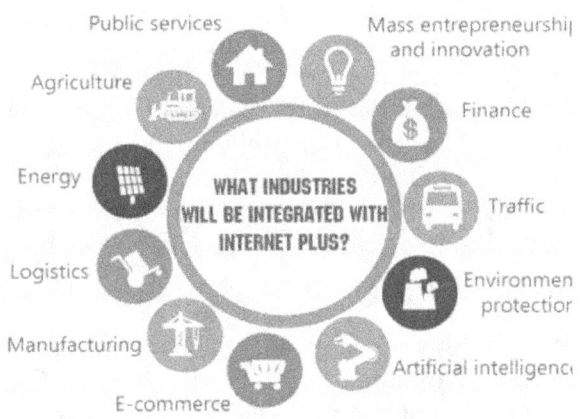

Figure 10-2 China's Internet Plus initiative

China's startups and tech giants pulled in a record US$58.8 billion from investors in 2017. The top investments went into logistics and transportation, US$16.8 billion, e-Ecommerce, US$12.7 billion, Fintech, US$4.8 billion, and internet services, US$2.5 billion. The largest single investment, worth US$5.5 billion, went to Didi Chuxing, which raised a total of US$9.5 billion in 2017. Didi Chuxing is more than a ride-hailing service. It also invests heavily in AI, a self-driving car research.

Alibaba accelerated its drive into brick-and-mortar stores by opening a supermarket chain, Hema. When a customer shops at a Hema store, their preferences are saved in its app. That makes it easier to order online and get deliveries if you prefer to do it that way next time.

Cainiao, a partnership company created by Alibaba in 2013 with 15 delivery companies, dedicates in e-commerce logistics, received US$799 million funding. Mobike, a bicycle sharing company, raised US$815 million. In April 2018, Meituan-Dianping, China's largest on- demand online service company,

Other blockchain applications

bought Mobike for US$2.7 billion. [322] Ele.me, an online food delivery company, raised US$1 billion. Ele.me has 260 million users across 2,000 Chinese cities, ordering meals from 1.3 million restaurants. Also in April 2018, Alibaba bought controlling shares of Ele.me at its valuation of US$9.5 billion.[323]

Bytedance, the new media firm behind news app Toutiao has its valuation jumped to more than $20 billion. Toutiao alone claims 120 million users. Pulling together news from around from other sources, Toutiao has 120 million readers each day who spend an average 74 minutes flicking through the app – double the time spent on Snapchat. Toutiao's popularity derives from the use of machine learning to create customized newsfeeds based on in-app behavior and reading preferences, similar to what Cambridge Analytica is doing. In 2017, it spent US$800 to acquire video-based social app – Musical.ly, another Chinese startup. Musical.ly has captured over 90 million users. It is a hit with North American teens, rising to the top of app store rankings next to Instagram and Snapchat. People can upload 15-second videos of themselves singing popular songs, dancing, or performing comedy skits. Some Musical.ly users even have found fame through the app, launching their own clothing lines.

Meituan-Dianping, which raised US$4.4 billion in 2017, is the Chinese equivalent of Groupon. By leveraging AI technology and innovation within its platform strategy, Meituan-Dianping is empowering local businesses to improve their business performance, and transforming the whole service industry into a digital ecosystem.

Section 10.01 Major Fintech companies in China

---

[322] "China's Meituan Dianping acquires bike-sharing firm Mobike", CNBC, https://www.cnbc.com/2018/04/04/meituan-dianping-acquires-bike-sharing-firm-mobike.html

[323] "Alibaba Takes Control of Ele.me, at $9.5 Billion Value", Shelly Banjo et. al., https://www.bloomberg.com/news/articles/2018-04-02/alibaba-buys-ele-me-in-deal-that-implies-9-5b-enterprise-value

Bitcoin, Blockchain & Fintech

China's financial services sector is ripe for change, exuberated by exponential growth in digital connectivity; deep penetration of smartphones; the explosion of e-commerce and a core of restless internet giants.

Chinese Fintech market can fall into the following categories: electronic payments, online financing, supply chain management, consumer financing and P2P lending, e-insurance, online fund and personal finance management, and online brokerage.

Tencent, a leading value-added internet service company in China, entered the Fintech arena with a simple social media app called WeChat. It was originally only a messaging app. However, it morphed into a huge Fintech app when it developed WeChat Pay function. By 2017, it claims a whopping 1 billion users.[324]

In 2014 Chinese New Year, when Tencent introduced the Hongbao (red envelope) function in the WeChat Pay, it became an instant success. Two years later, Tencent's WeChat Pay platform handled 32 billion digital Hongbao transactions over the six-day Chinese New Year holiday. This compares to 4.9 billion transactions PayPal did for the entire year of 2015.

Tencent operates more than a dozen data centers across China. It is moving stealthily to open overseas data centers.[325] Its Silicon Valley data center opened in April 2017, followed by four more in Frankfurt, Mumbai, Seoul, and Moscow by the end of 2017 to cater to its expanding client base. Its plan is to have eight overseas data centers in total to serve online games, online finance, video and other Internet-related industries. The new overseas data centers, located strategically in regional business or technology hubs worldwide, are part of Tencent's broader strategy to invest in the latest technologies, such as cloud, security, Big Data and artificial intelligence, to capture the next wave of

---

[324] https://www.statista.com/statistics/255778/number-of-active-wechat-messenger-accounts/

[325] " Tencent Cloud to Open Five More Overseas Data Centers in 2017", https://www.prnewswire.com/news-releases/tencent-cloud-to-open-five-more-overseas-data-centers-in-2017-300445849.html

Other blockchain applications

growth. It is also to serve its Chinese enterprise partners to deploy their services globally as well as international companies expanding their businesses in China or other parts of the world. The company's cloud revenue has already more than tripled year-on-year in 2016 due to the expansion.

Tencent, which expanded its popular WeChat messaging app into an ecosystem of Online-to-Offline (O2O) services, over the past few years, pulled in users with its own web bank and its mobile payments platform WeChat Pay. It also joined China Rapid Finance, a consumer-lending marketplace, to offer other investment products.

Alibaba pursues the same strategy. It is expanding cloud services overseas. Alibaba's cloud sales soared 126.5% to US$675 million from 2015 to 2016. As a comparison, the size of the world leading cloud service operators is US$500 million of Google, US$1.6 billion of Microsoft Azure, and US$9.8 billion of Amazon. [326] Alibaba's Fintech subsidiary, Ant Financial, claims that over 100 million users have taken out loans. Ant Financial managed US$96 billion for 295 million clients by mid-2016.

Baidu's Financial Services Group (FSG) had US$3.7 billion in assets in 2017, which accounted for 12% of Baidu's total assets. FGS sells wealth management products and loans and offers web bank for customers. Those services are tied to its mobile payment platform Baidu Wallet with 100 million activated accounts at the end of 2016, 88% growth from 2015. Baidu also invested in Zest Finance, a U.S. Fintech company that uses Big Data for credit scoring purposes.

JD.com, one of the two largest online retailers in China, a Fortune Global 500 company, spun off its Fintech arm, JD

---

[326] "Guess Who's King of Cloud Revenue Growth? It's Not Amazon or Microsoft", Barb Darrow, http://fortune.com/2017/09/27/cloud-computing-revenue-growth/

Finance.[327] At the end of 2017, its e-commerce had 266.3 million active customers. Despite it is a retail company, JD.com focuses on the technology of logistics and developed the world largest drone delivery infrastructure. Its delivery system deploys AI, autonomous technology, and robots. It has recently started testing robotic delivery services and building drone delivery airports, as well as operating driverless delivery by unveiling its first autonomous truck. JD.com also has launched a food supply tracking system using blockchain in Beijing supermarkets and online stores.

Wanxiang Blockchain Labs, a subsidiary of Wanxiang Group, one of the world largest automotive parts makers, working with Ethereum, is the largest blockchain development backer in China. After purchasing 500,000 ETH tokens in 2016, they pledged $30 billion for the development of a smart city in Hangzhou - a nine square kilometer plot of land with a planned population of 90,000 people. In such a smart city, blockchains would power everything from the city's electricity grid to its traffic control system.

Wanda Group, the world's largest private property developer, operating hotels, theme parks, shopping malls, theatres, media companies and department stores, is also diversifying into Fintech. In 2016, Wanda joined Hyperledger, intending to build smart apps for its businesses. Soon after, Wanda announced its blockchain platform Polaris, an open source platform, for smart supply chains, pharmaceutical management, finance and invoicing. The company is also working with China's Ministry of Industry and Information to draft domestic blockchain standards.

There are also many other Fintech companies specialized in the certain area of financial services. To mention a few: Wecash[328][329] is China's first Internet credit assessment company

---

[327] "China's JD.com isn't in a rush to list its finance arm", Christine Tan, et. al., https://www.cnbc.com/2017/09/03/jd-com-ceo-richard-liu-on-a-future-jd-finance-ipo-and-beating-alibaba.html

[328] http://wecashgroup.com/

Other blockchain applications

that provides solutions for technology companies. FangDD offers an online platform through which homebuyers and home sellers can directly connect to each other. Jimubox is a Beijing-based marketplace-lending platform that provides small and medium enterprise loans and individual consumption loans to under-banked Chinese borrowers.

Such an interest is not limited to the financial or retail companies. Chinese manufacturing companies also see an opportunity to unify the fragmented data flows flowing through their large and complicated factory floors and supply chains by marrying a blockchain data layer with IoT devices. By doing so, they are bringing blockchain to the manufacturing floor, adding dimension to the existing China Manufacturing 2025 initiative, the Chinese version of Industrie 4.0. Although Industrie 4.0 is different from Fintech by targeting industrial applications rather than financial services, many of the basic technologies, such as IoT, Big Data, AI, communication etc. are the same. The development of one will help the development of the other. There are also sectors of the market that these two initiatives overlap, such as the supply chain management.

A point in the case, Taiwan's Foxconn, the world's largest contract manufacturers of electronics and best known for manufacturing Apple's products, is developing a highly efficient supply chain that could also track delivered goods.[330] Foxconn sees blockchain as a way for its suppliers to get easy financing. It formed a subsidiary called FnConn[331] in 2013 to provide loans and financing solutions to suppliers, including Foxconn's upstream and downstream supply chain. The company also supports other small

---

[329] "Wecash Completes US$80 Million Series C Financing Led by China Merchants Group, Forebright Capital and SIG Ventures", https://www.prnewswire.com/news-releases/wecash-completes-us80-million-series-c-financing-led-by-china-merchants-group-forebright-capital-and-sig-ventures-300443456.html

[330] "Foxconn reveals plan for blockchain supply chain domination", Michael del Castillo, https://www.coindesk.com/foxconn-wants-take-global-supply-chain-blockchain/

[331] www.fnconn.com

and medium-enterprise (SME) customers in the computer, communication, and consumer electronics (3C) electronic manufacturing industries. FnConn works to improve supply chain finance, support industry, and economic development, and drive better integration across the financial services industry in China.

Section 10.02   Chinese Fintech landscape

Developing a Fintech sector requires a highly skilled workforce with an open mindset. China's major technology hubs are located in major metropolitans, such as Beijing, Shenzhen, Shanghai, and Hangzhou, where the skilled workforce and top university graduates are available abundantly and the global connections are one hop away from international airports. Beijing hosts country headquarters of many technology multinationals both foreign and domestic, such as Microsoft, IBM, Cisco, JD.com, and Baidu. Shenzhen is home to tech giants such as Huawei, ZTE, and Tencent. Shanghai, an international financial hub, and Hangzhou is where Alibaba headquarter resides.

Fintech firms are also located near some of the world's top-ranked universities for technology and engineering, such as Peking University and Tsinghua University. These educational powerhouses create an unparalleled and constant pool of new talents for the tech industry. These regions are continuously investing to nurture a conducive environment for Fintech firms in China.

Chinese Fintech market was a product of the unique combination of several factors. In 2016, the Chinese online retail sales were approximately US$900 billion, 26% increase year-on-year with one-third of the population ordering online at least once a week. E-commerce is about 16% of the country's total retail sales. At the same time, China has an untapped market of underserved customers. Chinese bank's priority is to serve state-owned enterprise (SOE) first. Small-to-medium-sized enterprises (SMEs) normally have disadvantages to secure needed financial services, such as loan, investment, insurance and others, from banks, so that they turn to alternative providers, which provide service at higher cost. The rise of Fintech services caters to this huge market segment.

Other blockchain applications

Chinese start-ups also enjoy a range of benefits from the Government, such as eligibility for annual tax deductions. Those qualifying as New High Tech Enterprises (NHTE) pay taxes at a lower corporate rate of 15% compared to the normal 25%. Even Fintech firms already financially backed by large internet players have been able to secure funding from government-affiliated firms. For instance, when Alibaba's Fintech affiliate Ant Financial raised US$4.5 billion in a Series B funding round in April 2016, it attracted investments from state-controlled entities like China Investment Corp, CCB Trust, China Life, China Post Group and China Development Bank Capital.

Competition is heating up for funding Chinese start-ups. This led to a 252% expansion in Fintech investments in the 12 months ending June 2016. Almost all the investment funds are from domestic sources.

In May 2016, China saw its first Fintech angel fund, FinPlus Fund, [332] [333] backed by the Fugel Group. The angel fund intends to invest US$0.15- 0.75 million in seed funding into individual start-ups while providing corporate offices and access to Fugel Group's clientele. Other private investors include Chinese leasing provider, Bohai Leasing, which took part in a US$207 million Series C funding round in 2015 in Dianrong.com.

Some foreign investors are venturing in as well. Foreign venture capital firms, such as Silicon Valley's Sequoia Capital, launched a China division in 2005.

Although China has less extensive physical banking infrastructure compared to the US and Europe, its digital infrastructure is quite mature. Its online penetration rate is 54% as compared to the 89% in North America and 73% in Europe.[334] However, because of its large population, China has 750 million internet users of which 96% are mobile internet users. The smartphone is the universal internet access device in China. The accessible wireless networks in public in China's major cities are

---

[332] https://finplusgroup.com/
[333] Finplus, http://www.finplus.me/
[334] "White paper: China internet statistic",
https://www.chinainternetwatch.com/whitepaper/china-internet-statistics/

much more extensive than in the West. As a result, mobile online payment users reached 358 million by the end of 2015, with an annual growth rate of 65%. 58% of the population is using their smartphone to conduct financial transactions primarily through Alibaba's Alipay or WeChat Pay payment service. In Q2, 2016 alone, there were 6.3 billion mobile payment transactions.

The leading e-commerce companies compete with each other to offer more and better services to their customers by digging into consumption ecosystem. Such an ecosystem contains both financial and retail activity data, like a treasure cove, that provides valuable insights into the market as well as technologies needed.

In 2016, China's e-commerce sales, US$900 billion, were 47% of global digital retail sales. Online purchases in China will soon reach 18% of the country's total retail sales value, compared with 8% in the US, 12% in South Korea and 16% in the UK.

By 2020, mobile transactions will rise from 56% of all e-commerce sales to 68%. With Chinese government encouraging citizens to shop online, and mobile broadband subscriptions projected to reach 90% by 2020, the adoption of e-commerce activities can go beyond urban metropolitan into rural areas.

Having domestic dominance with high levels of internet and mobile penetration, China has the world's largest and most developed retail e-commerce market. It has created a unique, dynamic and rapidly evolving ecosystem, which is a fertile ground for innovation in commerce, banking, and financial services. Such an ecosystem has drawn many technologies into its development whirlpools, such as Big Data and AI.

For the year ending in June 2016, Chinese Fintech investments surged to US$8.8 billion, the largest of global investment in this sector. As this exponential growth continues, China's Fintech is evolving into a completely new digital financial marketplace. The digitally perceptive Chinese consumers are ready to embrace Fintech offerings, creating opportunities for Fintech firms willing to take on digital transformation.

The proliferation of internet, communication and blockchain technology creates new business models that did not exist before. The new business models aim to tap the US$27

Other blockchain applications

trillion personal savings of the Chinese consumers. Fintech firms intend to channel such pool of savings into wealth management, insurance, and private banking.

To meet these needs, the Chinese technology giants are aggressively creating innovative platforms with both financial and non-financial solutions for their customers. They are also investing heavily in emerging technologies to support next-generation financial services, such as blockchain and artificial intelligence.

The competition among the e-commerce players to lure online customers is fierce. Over the past few years, the battle has spilled over to the Fintech industry, with Baidu, Tencent, and several other tech giants offering short-term loans and wealth-management products through digital channels.

The major Chinese Fintech firms are aggressively creating all-inclusive platforms. Ant Financial, for instance, offers customers with multiple products with a package deal, including not only financial services but also other types of services such as transportation, dining, medical services, and much more. This has already reaped initial successes as measured by speed of customer acquisition.

More Chinese financial services players are developing their own digital and Fintech technology, launching e-commerce platforms, e-payment tools and online financing and wealth management services. The barriers are getting higher for the newcomers. Not only they have to leverage new technologies, but also to amass a huge user base and merchant information, the online banking capabilities, as well as the presence of a large number of brick and mortar outlets in order to survive. The ones with only part of the requirements must seek partners to collaborate, to acquire or to be acquired.

Nevertheless, traditional banks are not out of the market. New regulations require in-person verification to open accounts. Online-only banks need to collaborate with traditional banks to use their established infrastructure.

Domestic state-owned and private investors almost funded all of the Fintech activities in China. So far, foreign entrants have met with varying levels of success – limited by government restrictions on operations and investment, as well as cultural differences. In late 2017, China opened up the financial market to the foreign institutions. [335] This is a positive development for China and global financial firms alike. It brings additional resources and technology to China and prepares Chinese economy for further globalization. It will also provide Chinese firms with new capabilities in areas such as risk management and banking system automation. During Jack Ma's visit to the US in June 2017, he envisioned SME's in the US selling directly to the Chinese consumers online, thus opening huge market access for SME's in the US.

Foreign payment processing companies such as VISA and MasterCard have had to work through UnionPay when processing local transactions. Since June 2016, China started to accept foreign payment processors for domestic card clearing businesses.

Having great success at home, Chinese Fintech giants are eyeing to go global both to diversify revenue streams and to reduce their dependence on the domestic market. Such a move is a giant step into an unchartered territory. Chinese companies cannot expect to replicate and export domestic business models abroad. The peculiar conditions at home making these companies successful in China may not exist in other countries. The key question is whether Chinese Fintech players will be able to adopt abroad and compete internationally. There are many entry barriers such as: to set up cross-the-boundary transaction processing capabilities; to overcome international payment barriers; to face different regulatory environments; to build trust, brand names, and reputation; and to prevail over cultural differences. In addition, they may not count on the government support or favorable regulatory environment outside of China. The overseas market may not have enough underserved SMEs and consumers. They need to adapt to local cultural norms and expectations. They will also need to compete directly with the local or international Fintech companies and financial powerhouses. The best and safest

---

[335] "China makes historic move to open market for financial firms", Bloomberg News, https://www.bloomberg.com/news/articles/2017-11-10/china-to-allow-foreign-firms-to-own-51-of-securities-ventures

approach to do so is to collaborate with and invest in overseas partners.

However, armed with enough capital, experience and huge market at home, Chinese Fintech firms will make inroad to the global market. Eventually, they will play an increasingly important role in the global Fintech collaborations and drive additional technological innovation. Their experience abroad, in turn, will allow them to develop the market at home further to bring the Chinese consumer contribution to GDP at least 10% above today's level. What is evident is that the next phase of development in Fintech innovation in China will have a major impact on global financial services in the future.

In Asia Pacific region outside of China, segments of China's Fintech industry like payments and insurance is already leading with non-banks commanding 35% or higher market share within the span of two years. The major companies making an inroad into Asia Pacific region are Ping An, which leads digital insurance and Alipay, TenPay and UnionPay for payments/ remittances.

Section 10.03 Chinese credit rating system

Most people do not appreciate the importance of credit rating system to the overall economy. At a personal level, if one does not have a good credit rating, he may have a difficult time to obtain a loan or even a credit card. Rating downgrade can negatively impact company finance. In a larger scale, the credit rating can have a detrimental impact on country's economy. During the East Asia economic crisis of 1997, many countries' economy suffered exacerbated blow due to the lowering of their credit rating by rating agencies.[336] These are some visible examples. However, the impact of the credit rating system is more than just when there is a financial problem. A good credit rating system works to accelerate the economy as enzymes in a

---

[336] "The Procyclical Role of Rating Agencies: Evidence from the East Asian Crisis", G. Ferri, et. al.
http://onlinelibrary.wiley.com/doi/10.1111/1468-0300.00016/full

biological body to act as catalysts to accelerate chemical reactions. Likewise, a bad credit rating system can do harm to the economy.

The flow of capital in an economy is like the flow of blood in a body. Billions of saving deposit in the banks channel into entities to produce economic activities. Credit rating can determine where these capitals will go. Entities with good credit score get capital with lower borrowing cost. However, it does not mean that these entities will produce more economic activities for the capital they receive than other entities with lower credit score, especially if the score is not accurate. Inaccurate credit rating system can reduce the capital efficiency and increase the overall business cost, just like bad infrastructures. On the other hand, a sound credit rating system can enhance the efficiency of resource allocation and promote economic growth.

Most of the consumer finance, such as credit card, is unsecured. Although the individual amount is small, the aggregated amount is large. For example, in 2016, the US credit card debt is US$1 trillion, [337] [338] while the total consumer debt is US$3.9 trillion, and it is heading higher.[339] The growth of the debt means the expansion of bad debts. How to establish and strengthen the control capacity and collection capacity is very important.

To finance such large consumer debt, but not to drive up the cost of financing is a major challenge. When the consumer debt demand increases, but the financing channels cannot meet the demand, financing cost will go up. To open up the financing channels, such as the issuance of bonds, asset securitization, and other financing means, maybe not as easy because each financing channel has different tolerance to the risk, and requires to accurately gauge the risk. This is where the credit rating comes in.

---

[337] "US credit card debt skyrockets, approaching $1 trillion", https://www.rt.com/business/335249-us-credit-card-debt/
[338] "2017 Credit Card Debt Study: Trends & Insights", https://wallethub.com/edu/credit-card-debt-study/24400/
[339] "Average Credit Card Debt in America: 2017 Facts & Figures", https://www.valuepenguin.com/average-credit-card-debt

Other blockchain applications

The construction of a sound credit system is far from easy. Current consumer finance company data and its data integration are insufficient to build a sound credit rating system because of the lack of data coverage, data quality, missing standards, missing data, and rapidly changing personal credit. Data collection and integration deviation can occur during the process of data separation and other issues.

In addition, an entity's creditability is more than just its credit history and related data. It is essential to establish an effective credit score model by associating the credit history to the risk. The credit risk also needs to be quantified to remove the bias associated with the subjectivity.

Traditionally, banks have their own credit score model. Banks need to build up its historical data and find suitable forecasting models. However, due to the immense scope of the credit rating, banks are not able to do it effectively and accurately. A complete and effective credit-scoring model requires the overall composition of the targeted entity as well as the circumstances when the data are collected. If the development of the credit system is not comprehensive enough, the risk and cost for the banks are high.

An ideal credit agency acquires data from multiple sources; builds massive data handling, mining, and processing capability; determines the reliability of each data source with the anti-fraud and filtering model; does data cross-validation. After acquiring enough data, the building of the credit score modeling is no less challenging, because the credit data are not always coherent, and consistent. Until now, the tools available to do data collection, mining and processing are limited.

Fintech can help credit rating in two aspects: first is the acquisition and processing the data and second is the creation of a comprehensive model by deploying Big Data technology and artificial intelligence. Fintech will play an important role in this arena in the future.

Compared to traditional credit companies such as FICO, Fintech companies with Big Data technology have the advantage in the breadth and depth of the data, can gather more dimensions of behavioral information, and can discern a previously undetected pattern, which will, in turn, give a more comprehensive rating.

With the Big Data mining and analysis techniques, the data collected have less asymmetry. Through Big Data collection and data mining, consumer finance companies can identify their customer needs, customer portraits. Thus, they can do more accurately target marketing and form consumer credit judgments. They can also correlate borrowers to different consumer finance platforms and determine customer's repayment intention and ability under different economic environment. A stress test will reveal customers' financial capability under difference circumstances, such as unemployment, having a health problem.

Applying artificial intelligence and machine learning, such as random forest, neural network, gradient boosted decision tree (GBDT), can improve the modeling accuracy. Credit scores based on the artificial intelligence can make credit management work more efficient, objective and focused. Such a score can determine the entity's creditworthiness under certain circumstance.

The current state of credit rating system in China is still immature and this immaturity has caused high social financing costs, lending efficiency and industry's ability to assess risks, inhibited the economic vitality. Both the government and the industry recognized its importance. They are devoting great effort to develop credit rating industry by using new technologies, such as Big Data and artificial intelligence in recent years.

In 2009, the China Banking Regulatory Commission (CBRC) issued guidelines for the supervision of internal credit rating system for credit risk of commercial banks. In 2013, the Chinese State Council issued the Regulations on the Administration of Credit Industry. By the end of 2014, the central bank credit center had already accessed 1811 data institutions and covered 350 million people and business entities. With the gradual establishment of modern credit system helped by Fintech, a credit rating industry is emerging.

In September 2015, the State Council issued the plan for Big Data Development. In a series of policy, regulation, and guidelines, the PBOC and CBRC encouraged financial institutions to innovate consumer credit products.

In 2016, the Chinese government is taking a consolidated approach to develop an extensive nation-wide Social Credit

Other blockchain applications

System (SCS). The SCS is akin to the combination of credit rating systems in the US – FICO, Vantage score, CE score, Moody rating for bonds, S&P credit rating, Better Business Bureau (BBB) and many others, and even more comprehensive. One difference is that in the US, the private companies provide scores, but in China, it is the government which provides SCS.

SCS focuses on honesty in government affairs, commercial integrity, societal integrity, and judicial credibility. The scope embraces both personal and business credit and trustworthiness rating. Such a rating can easily translate into "trust" for the blockchain validation. In an era when the daily business dealings involve totally unknown parties, such a trust system becomes essential. SCS is intending to remove the "Trust" barrier. However, there are critics condemning such a system is a Big Brother.[340] SCS will come online by 2020.

The Chinese credit rating industry is by no means limited to the government. The Government also has granted licenses to eight private enterprises to develop their credit scoring systems. Among the licensees are e-commerce companies Alibaba, Tencent and Ping An Insurance, and five credit rating companies: Pengyuan Credit Services, China Chengxin Credit Information, IntelliCredit, Credit Arm and Yin Zhi Jie.

Independently, in September 2016, China's National Internet Finance Association (NIFA) also launched its Internet Financial Industry Information Sharing Platform (IFIISP). NIFA, run by the central bank and with 400 member traditional financial and internet finance companies, including heavyweight companies such as Ant Finance, JD Finance, Lufax, and Yirendai, was established to regulate Chinese Fintech firms and control risks in the sector. The IFIISP provides credit data on a customer by sending requests to all other member companies and collating the results without divulging the source data to protect competitive insight. The NIFA took a giant step forward in developing a proper credit score system with IFIISP.

---

[340] "Big Data meets Big Brothers as China moves to rate its citizens", Rachael Botsman, http://www.wired.co.uk/article/chinese-government-social-credit-score-privacy-invasion

Even so, China's P2P regulations remain relatively less burdensome than those in developed markets do, whether by design or by omission. However, as the system evolves, the government sees greater need to tighten the regulations. In 2001, President Bush signed a Patriot Act, which required certain ID verification for the customers to open a bank account. In 2016, China followed the suit, with concerns over money laundering and fraud, to set up new regulations that required in-person verification to open a bank account. Since the online-only banks do not have a physical branch, they need to collaborate with traditional banks. Thus, traditional banks have gained an additional source of business from the online banks.

Chinese legislature is including a more substantive data protection framework. The new framework includes PBOC's additional requirements for non-bank payment institutions around effective protective measures, risk control systems and KYC measures, and storage of sensitive information. The Chinese government will likely continue to take a leadership role in setting the agenda on data privacy and protection as this will increasingly become a central pillar of the financial services marketplace – critical in enabling the Fintech sector to operate and prosper.

In addition, some private enterprises already moved ahead with their own versions in credit rating. As e-commerce companies already generated a huge amount of data from its own platforms about the customers, suppliers alike, they set up their own credit rating system. For example, the scoring platform called "Sesame Credit Management" developed by Alibaba's Ant Financial analyses data generated on Alibaba's shopping platforms from 300 million customers and 37 million small businesses. Customers with higher credit scores enjoy special privileges. Other similar systems built by Tencent using the data from its own social media app, WeChat and WeChat Pay. They also offer the credit rating services to other financial services providers. For example, Jubao uses social media data generated on WeChat and Weibo to assess customers' creditworthiness. Meanwhile, JD Finance is collaborating with US-based ZestFinance in a Joint Venture to develop services of credit risk evaluation and extend consumer access to credit in China.

Other blockchain applications

With such a torrent of activities, China is setting up world-class credit rating system quickly. In a few more years when these systems are up and running, people will feel more confident in transacting with unknown parties and pushes China's Fintech market even further. People will also be more cautious to do things that will spoil their credit rating. Credit rating system will support risk-adjusted loan growth to bolster consumer spending, further supporting the growth of Fintech.

Currently, China's consumer expenditure is only 39% of the GDP, vs. US 68 %, France 55%, and Germany 63%. [341] Stimulating household consumption is one way to boost GDP growth. Fintech will help.

Chinese Fintech firms are leveraging Big Data from e-commerce, messaging, search, social media and other internet-based services to personalize the customer experience, provide new services, and leverage operational efficiency. Customer data can be used to support other online revenue streams, such as lending, insurance, investment and wealth management.

Currently, the development of most of these systems is in progress in parallel. As China's mobile financial market matures, the interoperability across mobile wallets and other banking services will receive more importance to improve efficiencies, democratize the ecosystem and level the playing field between new entrants and incumbents. Such an open-architecture policy would start a race to offer the most innovative solutions, as players attempt to differentiate themselves from the pack.

Section 10.04   Chinese online lending market

One of the heated Fintech markets is the online lending. Since many people and SME's are unable to obtain loans from banks, the online lending demand is huge in China. These online lending Fintech companies are serving the same market segment as the German Fintech company: Kreditech.

---

[341] "Household final consumption expenditure (% of GDP)", World Bank, https://data.worldbank.org/indicator/NE.CON.PETC.ZS

SME's, which lack qualified collateral and track records of credit repayment, receive only 20-25% of bank-disbursed loans. Yet they account for 60% of GDP, 80% of urban employment, and contributing to 50% of fiscal and tax revenues in China. SME's also suffer from asymmetric information, with limited transparency in their financial positions and credit rating assessments. Even if they do secure bank loans, the interest rates are much higher than that of large corporations, based on the risk concerns.

Likewise, retail customers also receive lower priority at China's banks. This provides a fertile ground for the growth of non-bank operated online lending market. About 160 million people in China took out US$180 billion in online loans in 2016. There are many companies operating in the niche markets. For example, a Fintech online financing company, Daikuan, which raised US$2 billion in 2017, offers online loans for buyers of second-hand cars.

However, the rapidly growing market is loosely regulated, and plagued by fraud and defaults. The worst P2P lending fraud happened in 2014-2015,[342] with which 900,000 investors lost over US$7.6 billion in a Ponzi P2P lending scheme of Ezubao. As a result, people lost confidence in the P2P lending. Half a year after the exposure of the fraud, nearly one-third of all online lending companies in China was in financial trouble and 40% and 1,600 P2P lenders exited the market by April 2016.

The P2P market has cooled down and the surviving P2P lenders sought to collaborate with banks to restore their reputation and re-establish their credibility, even though they were not involved in the scam. For example, a collaboration between Dianrong.com and the regional Bank of Suzhou set up a P2P loans platform targeting SME's. The Shanghai-based P2P lender China Rapid Finance (CRF) collaborates with China Construction Bank to create a P2P platform providing investors access to CRF's P2P offerings via the bank. Other collaborations are between P2P lenders, such as Jimubox, RenRenDai, Minshengyidai, and China Minsheng Bank for the bank to manage and safeguard investors' funds. Even CreditEase, a Beijing based Fintech leading

---

[342] "Ezubao", https://en.wikipedia.org/wiki/Ezubao

conglomerate with a nationwide network in 255 cities in China, made a similar arrangement with China Citic Bank.[343]

Fintech experts believe that artificial intelligence and blockchain technology can prevent the fraud. It prompted many P2P lenders to sought blockchain solutions. To prevent the scam from happening again, the Chinese central bank PBOC also introduced new regulations to oversee P2P lending and online payments in China, such as the imposed credit limits, required a principal guaranteed by the platform, etc.

The Ezubao event is a temporary setback for the P2P lending. New regulations and platforms with new technologies are gradually restoring the consumer confidence in P2P lending. Once the confidence returns, the market is expected to grow at an annual rate of 50% again, because there is an undeniable demand.

Non-banks, which appeared in the scene in the last decade due to the flourishing e-commerce, have an upper hand in China especially when it comes to competitiveness in offerings, digital functionality and experience, quality, innovation, and even trust levels. For these reasons, SME's and retail consumers are increasingly turning to the non-traditional lenders for financial services.

The SME, which takes part in the supply chain of e-commerce, can get loans from the e-commerce company, which lends to SMEs by leveraging SME's merchandises on its platform. There is no risk to them because once the SME products are sold on their platforms; they are the ones who receive customer payments. Key participants include Ant Financial and Alibaba's MyBank, WeChat's WeBank, JD.com's JD Finance and Gome Electronic Appliance.

Even the non-e-commerce P2P lenders are catering to individuals and SME's customers. Market leaders are Lufax of Ping An Insurance, Yirendai of CreditEase, Rendai, Zhai Cai Bao of Alibaba and Dianrong. The risk for these lenders, without the leverage of e-commerce platform, is higher; therefore, the interest rates are also higher. But, their service opens to anyone, not

---

[343] http://english.creditease.cn/index.html

limited to the supply chain participants. Lufax had more than 23.3 million users by June 2016, doubled in one year.

In the CreditEase platform, for example, one can apply to post as either a borrower or a lender. It is like an auction market of the loan. Once the condition matches, the lender grants the loan to the borrower, and the borrower pays back the lender in monthly installments with interest. Most often, lender's money is dispersed into many borrowers, and the borrower gets a loan from multiple lenders. Such a diversification reduces the risk. In case, the borrower misses the payment, the lender will receive the payment from CreditEase, which will collect from the borrower. CreditEase is acting more than a matchmaker but a guarantor of the loan. Therefore, CreditEase, not the lenders, carries the default risk.

The lender commits funds, which can be automatically allocated among approved borrowers or the borrowers at his choice. The lender, also known as an investor in the platform, can choose to enroll in the automated investing option, which automatically reinvests investors' funds when he receives payments from the borrower. Alternatively, the lender can select lending opportunities to the approved borrowers by lender himself. Furthermore, the investor's fund is not locked to its maturity. He has the option to sell his loan in the company operated secondary loan market at any time. This liquidity offers investors the opportunity to enter and exit their investments without waiting until maturity.

There are also other types of P2P lending. Using a different business model, China Rapid Finance (CRF) is China's largest consumer lending marketplaces in terms of a total number of loans granted. Much like the e-commerce marketplace, where the platform operator offers online spaces for buyers, CRF operates the marketplace without taking credit risk by using machine learning and proprietary technology to select qualified borrowers. For the first time borrowers, the amount of loan is limited. After the customers establish their credentials, the loan amount

Other blockchain applications

increases. Its board of directors includes reputable global financial executives with extensive experience.[344]

CBRC, China Banking Regulatory Commission, newly proposed restrictions for banks would pave the way for P2P lenders to enter WM and deliver financial advice in an innovative way. After the Ezubao fraud, smaller P2P lenders exited the market. That leaves the larger, more established online lenders to dominate the largest online lending marketplace in the world.

Chinese Fintech firms are increasingly to look for technology solutions to enhance security and to reduce risk. In March 2017, CreditEase launched Ethereum based blockchain service called Blockworm to enhance its security.[345] It is expected that when the Social Credit System (SCS) is deployed, P2P lending can receive another boost in consumers' confidence.

Section 10.05 Chinese wealth management, e-insurance, and online trading

Until now, the Chinese investors lack investment choices. Many of them choose to park their money in real estate. This has driven up the housing price and caused a real estate bubble. Many houses sit unoccupied, while many people cannot afford to buy a house. Wealth Management (WM) carters this group of growing middle class with huge savings looking for better returns. The online fund management offers an outlet for the savers to receive a better return than bank's low-interest rate. The value of wealth management products (WMPs) expanded 56% to US$3.5 trillion in 2015. The online fund management app linked to payment platforms offers ease of access. The investment opportunity is never a few clicks away on the smartphone.

Primary participants are Yu'e Bao of Ant Financial, LiCaiTong (Tencent), and Baifu (Baidu). Alibaba's Yu'e Bao is

---

[344] China Rapid Finance website, http://stage.investorroom.com/chinarapidfinance/index.php?s=118

[345] "FinTech CreditEase Launches Ethereum-based Blockchain Service", Cindy23, http://news.8btc.com/fintech-creditease-launches-ethereum-based-blockchain-service

one of the world's top money market funds with US$170 billion of assets in 2017.[346] Licaitong is Tencent's wealth management platform accessible via WeChat wallet and QQ wallet. It provides Tencent users with quality investment selections in a transparent and user-friendly investment environment. Most of these products offer a yearly return of 4.5% to 5.2% depending on the investment duration and an amount up to US$1.5 million. In 2016, the total transaction amount reached US$300 billion. China also houses several of the largest mutual funds, such as Yu'e Bao and Tencent's Licaitong.

Another growing market is the e-insurance. The insurance premium via online distribution channel grew at a whopping 235% to US$34 billion from 2014 to 2015, as compared to 11% to US$331 billion via the traditional channel.[347] It still has a huge room to grow. An increasingly urbanized and well-educated population, with rising household incomes and personal financial assets drive life insurance in China. China' auto insurance market is already the second largest in the world, but the penetration rate is still behind Western countries. Other insurances such as home insurance and business insurance are also growing.

Traditional Chinese insurers are scaling up their own digital expertise. More than 100 out of around 130 traditional insurers have introduced online sales platforms, distributing US$30 billion more premiums from 2014 to 2015, equivalent to 9% of aggregated premium value.

As traditional insurance companies have strengthened their online presence, it has become increasingly difficult for Fintech firms to penetrate this market without collaborating with existing players. Therefore, most Fintech firms seek to collaborate with the existing insurers. By doing so, Fintech firms benefit from traditional insurers' experience in navigating regulatory hurdles, tapping into risk assessment, pricing analytics, and other technical

---

[346] "Alibaba's Yuebao becomes world's largest money market fund", Yang Jing, https://news.cgtn.com/news/3d41544f79637a4d/share_p.html
[347] "China's insurance market", whitepaper, http://www.aon.com/inpoint/bin/pdfs/white-papers/ChinaWhitepaper2016.pdf

Other blockchain applications

expertise. Traditional insurers benefit from Fintech's new channels. Despite its increasing digital capability, the insurance industry remains receptive to Fintech collaboration. Fintech companies offer a huge database of potential customers, enabling traditional insurers to tailor solutions for customers with different risk profiles at competitive premiums.

China's first online-only insurer, Zhong An, was launched in 2013 as a collaboration between Alibaba, Tencent, and China's second-largest insurance firm, Ping An. ZhongAn has a technology development arm to develop technology based on artificial intelligence and big data to simplify insurance.

To develop the Fintech products, Baidu formed a joint venture with international insurance giant Allianz and Hillhouse Capital and CPPIC, using Big Data and machine learning for better risk assessment in e-motor insurance. Some Fintech players go for acquisition and merger. For example, Ant Financial took a 60% stake in Cathay Insurance, the Chinese insurance unit of Taiwanese Cathay Financial Holdings. Alibaba broadened its insurance product offering and made available on Leyebao, its online insurance sales platform while creating insurance products to suit the needs of SMEs operating on Taobao.

In China, e-insurance companies sell insurance policies through e-commerce and online wealth management (WM) platforms. Major players are the People's Insurance Company of China (PICC), Ping An, and Zhong An. Many startups are developing the blockchain solutions for insurance. The most notable one is Chain B,[348] which is a blockchain startup to target the insurance industry. It aims to develop a decentralized network to provide multiple validation points for claims without the costs associated with traditional insurers.

The action is not limited to startups; China's big banks are also launching blockchain applications aimed at insurance

---

[348] "Digital insurance in action", The digital insurer, https://www.the-digital-insurer.com/dia/chain-b-blockchain-enabled-insurance/

sector.[349] For example, China Construction Bank (CCB) began using a custom blockchain platform, jointly developed with IBM, in Q3, 2017, to sell third-party insurance products to a distributed ledger.

The Shanghai Insurance Exchange had conducted a blockchain trial on insurance businesses with nine large insurance companies, including Cathay Life Insurance, Meiji Yasuda Life Insurance, AIA Group, and others in China in 2017.[350] The platform, in collaboration with IBM, integrates several new technologies including blockchain, Big Data, biological recognition and artificial intelligence.

Online stock and mutual fund trading is another big market. Players include Ant Financial (Alibaba), LiCaiTong (Tencent), Baifa (Baidu), Wacai, Tongbanjie, JD Finance (JD.com) and other online brokerage, social network and information portals, such as Snowball Finance, Xianrenzhang, Yiqiniu, Tiger Broker (backed by China's smartphone maker Xiaomi) and Futu Securities (funded by Tencent). All of them have mobile apps for stock and mutual fund trading. By mid-2016, Yu'e Bao had more than 295 million customers, making it one of the largest global consumer funds.

Most of the Chinese P2P lenders such as RenRenDai and CreditEase are also WM service providers. With their access to massive social and analytical data around customers' creditworthiness and purchasing trends, these Fintech firms have opportunities to build advanced finance platforms to support elevated expectations from increasingly demanding investors.

In mid-2016, WM firms introduced simplified investment advice service using robo-advisors through sophisticated automated online platforms incorporating Big Data and artificial intelligence. The robo-advisor uses modern portfolio theory, Big Data

---

[349] "Big Four' Chinese Bank to Launch Blockchain Bancassurance Product", https://www.coindesk.com/big-four-chinese-bank-launch-blockchain-bancassurance-product/

[350] "10 insurance firms test blockchain for insurance in China", Wolfie Zhao, https://www.coindesk.com/insurance-firms-blockchain-test-insurance-china/

Other blockchain applications

algorithms, XUANJI's asset allocation solution, quantitative modeling and program trading using machine learning. The result is substantially lower costs to provide universal access to entry-level, affordable, yet customized online financial advisory services.

One of the top robo-advisory providers is CreditEase, which launched its robo-advisory product ToumiRA to provide cost-effective access to international WMPs for Chinese retail investors, using trading algorithms to match investor risk preferences and objectives to their optimal portfolio. Another robo-advisory platform Xuanji, launched by PINTEC, provides intelligent investment advice to retail investors with the ability to customize and automatically rebalance a global portfolio.[351] In addition, PINTEC also offers a B2B version for brokers, independent financial advisors, and financial institutions. Both versions can trade USD ETFs and RMB denominated Chinese mutual funds.

Others offering of the machine-assisted investment advisory includes Baidu Gupiao, PingAn One, MiCai and Clipper Advisor. Robo-advisory could well reshape the future of China's WM business. To survive such competitive market, they will need a breadth and depth of asset offerings, portfolio allocation, and technological superiority in Big Data analytics and machine learning.

Many Chinese Fintech companies also form alliances to leverage each other's strength in technology, marketing or database. For example, China Ledger Alliance, comprising regional exchanges, created an open source blockchain protocol to support IoT development. Financial Blockchain Shenzhen Consortium, including members such as Ping An Insurance, a member of global consortium R3 and Tencent, collaborates on research and group-wide blockchain projects, with a focus on capital markets technology, securities exchange, trading platforms, banking and life insurance. It aims to create a securities trading

---

[351] "Chinese robo-advisor Xuanji launched by PINTEC group in Beijing", Steven Hatzakis,
https://www.financemagnates.com/fintech/investing/chinese-robo-advisor-xuanji-launched-by-pintec-group-in-beijing/

platform prototype and develop credit, digital asset registry, and invoice management services.

Qianhai International Blockchain Ecosphere Alliance aims to establish an efficient ecosystem for developing blockchain technology and its applications. The Alliance membership includes world leading global technology companies such as Microsoft, IBM and Hong Kong's Applied Science and Technology Research Institute (ASTRI). It hopes to accelerate the commercialization of blockchain R&D and promote its application to support China's social and economic development.

Aggressive Chinese tech firms equipped with skilled IT infrastructure and application development are looking to go global. They are investing in global technology hubs, penetrating new product markets overseas to diversify revenue streams and reduce domestic reliance. Baidu, Alibaba, and Tencent are expanding to serve outbound Chinese travelers abroad and expats, while seeking out new customers in emerging economies from Africa to South and South-east Asia. They are particularly interested in the future possibilities of new O2O revenue streams. For example, Alibaba currently has more than 86% of revenues from China but aims to generate half of all sales from overseas. The Group is developing an international ecosystem that encompasses targeted marketing, logistics, payment services and cloud computing.

Their strategy is to achieve globalization via international acquisitions and expansion. For instance, its AliExpress e-commerce platform is already doing a brisk business in markets such as Russia and Brazil. Alibaba made its largest international investment to-date in April 2016 with US$1 billion for the control of Lazada Group, Southeast Asia's largest clothing and electronics portal. In the summer of 2017, it invested additional US$1 billion and in March 2018, it has doubled down on Lazada with additional US$2 billion investment with a total investment of US$4 billion.[352] To further increase exposure into Southeast Asia,

---

[352] "Alibaba doubles down on Lazada with fresh US$2 B investment and CEO", Jon Russel, https://techcrunch.com/2018/03/18/alibaba-doubles-down-on-lazada/

Other blockchain applications

Ant Financial took a stake in Ascend Money, a Thai online-payment provider.

In 2015, Ant Financial ventured into India with a US$680 million purchase for about 40% of Paytm, India's largest mobile commerce platform with 122 million users and 23 million mobile wallet users, and another US$100 million for online marketplace Snapdeal. In 2017, Alibaba has invested additional US$177 million in Paytm E-commerce, a spinoff of Paytm.[353] These investments gave Alibaba a payments banking license in India and an immediate foothold in a country with an exponential growth in online payment industry of CAGR of 50% from 2007-2014.

To expand the business, Alipay has announced ambitions for one million offline partner merchants globally within three years, a move that will allow 120 million Chinese tourists traveling abroad every year to pay with Alipay even when they are abroad. The first of these cross-border payments partnerships was rolled out in 2015, allowing Chinese tourists with an Alipay account to shop and pay at 70,000 overseas merchants with the app. With 450 million active registered users and 200 financial institution partners, Alipay is collaborating with leading global payment providers to ensure international merchants can be integrated into handling the Chinese payment platform at home.

Ant Financial and European retail merchant leaders, Germany's Wirecard and Concardis, established a partnership to serve the growing number of Chinese tourists in Europe. Alipay also entered an agreement with Ingenico to embed Alipay into Ingenico's payment portal, the largest in Europe. This allows European merchants to accept customers to use Alipay for their purchases. A global agreement with the insurance giant AXA allows Alibaba/Ant Financial to sell AXA travel insurance to outbound Chinese travelers.[354]

---

[353] "Alibaba to invest US$177 m in India's Paytm", Simon Mundy, https://www.ft.com/content/5cbb69bf-a2ae-3288-8500-27656a12067b

[354] "AXA, Alibaba and Ant Financial Services announce global strategic partnership", https://group.axa.com/en/newsroom/press-releases/axa-alibaba-ant-financial-services-announce-global-strategic-partnership

Likewise in the US, the partnerships with San Francisco-based ride-hailing service Uber Technologies, Airbnb and Macy's for Chinese customers to use their Alipay wallet. Baidu has also created its own alliance with overseas merchants, in Thailand, South Korea, Japan, Hong Kong, Macau, and Taiwan.

Section 10.06 Chinese third-party payment market

Mobile payment is a precursor of digital currency. In China, the non-bank third parties handle the payment from the buyer to seller. Thus, it earned the name of third-party payment. The third party payment system has its roots in the e-commerce, where the platform owner acts as an escrow. When the buyer pays, e-commerce company tells the seller to ship the product. When the buyer receives the product, e-commerce company releases the payment to the seller. The third party transfers the payment from buyer's bank account to seller's bank account. This method is necessary because few people in China had a credit card. For example, China had only 0.29 cards per capita at the end of 2015 in stark contrast to an average of 2.35 credit cards per capita in the US.[355]

Compared with traditional payment methods and credit card payment, third-party payment has its obvious advantages: convenience, cost saving, added security features, and accessibility. Certain unique features of China's e-commerce environment have further contributed to the popularity of e-commerce in China. The threshold for applying for credit cards from a bank is high in China. Even for those who have credit cards, many are wary of using them for online purchases for fear of leakage of personal information. Banks have also imposed stringent maximum transaction limits on online transactions effected through credit cards.

People have the impression that the third party payment is free of service fee. It is not exactly true, although its cost is indeed

---

[355] "Average Number of Credit Cards Per Person: 2017 Card Ownership Statistics", https://www.valuepenguin.com/average-number-credit-cards-per-person

Other blockchain applications

much lower than that of credit cards. When the party works with banks, banks charge transaction fees. Therefore, Tencent and Alipay alike, which used to provide payment transfer, free of charge, now started charging customers a small fee. Such a fee is still relatively inexpensive: 0.1% fees for transaction amount over US$3,000.

The dominant domestic online marketplaces for consumers are Taobao (Alibaba), Tmall and JD.com, which handle payment transactions between buyers and sellers as third parties. As the e-commerce proliferated, the third-party payment platforms also grew. At the same time, as the mobile phone penetration got deeper, mobile payment apps started to appear to make the payments easier. Soon the convenience of this third party payment system migrated from e-commerce to the physical stores. Today, most of the brick and mortar stores, street vendors, mom and pop shops, supermarkets, restaurants also joined the mobile payment system.

In a nation where cash historically dominates the consumer market, China is moving from cash to digital payment on smartphones, bypassing credit/debit cards. Fintech and blockchain technology can greatly accelerate the use of payment platform and enhance its security.

In the US and Europe, Fintech has been driven by startups or financial institutions. On the other hand, China's internet giants have largely been the sources of capital for its Fintech firms. Ant Financial, which operates the Alipay payment platform, was a spin-off from Alibaba.

Chinese third-party payment market in 2016 reached US$5.5 trillion, which is roughly the same size as the US credit/debit card payment market. According to the "2016, The Federal Reserve Payments Study", [356] the US credit/debit card transaction volume is US$3.16 trillion and US$2.56 trillion, respectively, totaling US$5.72 trillion, while the US mobile payment volume is around US$112 billion. The rapid spread of mobile payment platforms in China are pioneered by Alipay,

---

[356] "Federal Reserve payments study 2016", https://www.federalreserve.gov/newsevents/press/other/2016-payments-study-20161222.pdf

TenPay, JD Pay and UnionPay's Quick, ICBC's e-wallet, 99bill, and others.

The earliest entry was Alibaba, which launched Alipay in 2004 to serve its Taobao e-commerce platform. Alibaba's Alipay was the most used third-party payment platform. In part, the Chinese flourishing e-commerce market and its adoption of internet and mobile payments can also be due to the presence of a massive domestic retail market.

Alibaba's Alipay is now the largest online payment gateway in China, accounting for half of the Chinese third-party online payments; Tencent's TenPay currently ranks second. Apple also launched Apple Pay in China in February 2016.

These e-commerce firms were not content at providing the payment service only. The fiercely competitive market serves as a driving force to create new and innovative market applications. They obligate to act under market force or foresee big opportunities in using similar technology to create new markets. Soon, they used third-party payment platform as a springboard to enter other Fintech business, such as online lending, e-insurance, credit rating, wealth management, stock trading, bike sharing, and many others. Once it started, it spread like a wildfire to all the possible imaginable applications. This is how the internet and e-commerce giants — Tencent, Alibaba, JD.com entered the arena of Fintech.

Alibaba spanned off the payment service branch as Ant Financial, which manages Alipay. Alipay works with the traditional banks, credit card companies, including VISA and MasterCard.

Merchants and customers using Alipay could easily park their excess cash in Yu'e Bao to earn an attractive interest that banks were unable to offer. This natural extension of Alipay's payment service to money market fund service resulted in exponential growth. In June 2013, Alibaba launched money market fund Yu'e Bao, run by Tianhong Asset Management, another Alibaba affiliate. By the end of December 2017, Yu'e Bao's market fund assets reached US$233 billion, the largest in

Other blockchain applications

the world. Its close competitor is JP Morgan's US Government Fund at US$140 billion.[357]

In addition, to serving its e-commerce platform, Alipay also expanded to more than 460,000 Chinese businesses. Recently, it is making inroads into the international markets by signing up overseas merchants and accepting foreign currencies. It has long dominated China's mobile payment market until its competitor Tencent came up with a more innovative mobile payment system, TenPay. Alipay saw its share of the market fell from 71% in 2015 to 54% by the end of 2016, while TenPay share rose from 16% to 37% during the same period.

In June 2015, following the success of Yu'e Bao, Ant Financial launched MYbank, a new online-only bank in China. The online-only bank does not have a physical branch office. All the transactions are online.

Chinese mobile device manufacturers, such as Huawei and Xiaomi are also moving into Fintech and mobile payment partnerships, e.g. UnionPay. The regulatory environment has been generally facilitative to the verticals collaboration –from e-commerce, gaming, chat and search engines to financial services. The offline-to-online interaction is also popular in China.

Tencent's meteoritic rise in the Fintech arena thanks to its social media app called WeChat. WeChat users are able to transfer money between each other, as well as pay for services such as taxis, digital subscription, food delivery and restaurant bills using an embedded function called WeChat Pay or TenPay.

TenPay takes advantage of the huge user base of the WeChat app ecosystem. With 890 million users as its customer base in the online messaging service, WeChat entered the mobile payment market a decade later. Anyone with a WeChat account can send and receive payments to anyone else with a WeChat

---

[357] "China's giant Yu'e Bao money market fund riskier than US rival, Fitch says", Gorgina Lee, http://www.scmp.com/business/money/markets-investing/article/2124465/chinas-giant-yue-bao-money-market-fund-riskier-us

account. The app allows users to keep funds in its wallet for peer-to-peer payments, in-app purchases. TenPay also has signed up physical stores, the off-line merchants, including Starbucks, which has 2,600 stores in China. By doing so, WeChat transformed itself from social media platform into a payment platform in 2013 and launched a personal online investment fund in January 2014. One year later, WeBank, China's first online-only bank also launched the same product.

WeChat Pay is targeting to cover more than 10 million small merchants or stores in China. It has an innovative function called the Digital Red Envelope, or Hongbao in Chinese. The traditional Hongbao is a red envelope containing gift money in cash that the elderly gives to young kids in Chinese New Year or a wedding gift. The ability to transfer money online via WeChat accounts revolutionized Hongbao tradition. Using WeChat's Hongbao function, one can send Hongbao electronically instead of face-to-face delivery. It became widely popular. In 2016, 64 billion Digital Red Envelops were exchanged over the six-day holiday period. An added benefit to the merchants, the mobile payment system collects the data gathered from spending habits and financial information. It, in turn, allows merchants to target their specific customers.

Facing threat from Tencent, Alibaba is not standing still. It is building its own physical network of stores, both domestically and overseas. It has already signed up more than 2 million brick and mortar shops in China with 10 million merchants on Taobao using Alipay. Ant Financial,[358] a spin-off of Alibaba, provides more financial services offerings than the money market fund giant Yu'e Bao to attract customers.

Apple launched Apple Pay in China in May 2016.[359] 30 million bank cards signed up on the first day. However, its market share did not grow as expected due to several reasons. First of all, Apple Pay is a latecomer. The mobile payment market was already well saturated and dominated by two big players with fierce competition. In addition, Apple Pay uses NFC (Near Field

---

[358] https://www.antfin.com/index.htm?locale=en_US
[359] "Apple pay is coming to China in 2016", Rich McCormick, https://www.antfin.com/index.htm?locale=en_US

Other blockchain applications

Communication) vs. the QR codes, used by both Alipay and TenPay, which introduced QR code system in 2011- 2012. Soon afterward, QR codes spread quickly in major Chinese cities.

There is an underlying reason that NFC cannot duplicate the success of QR code in China. NFC payment requires dedicated NFC-equipped smartphones and the point-of-sales terminals in the store. It is unreasonable to expect millions of Mom and Pop shops and street vendors to sign up for the NFC equipment, while one can easily print QR code on a piece of paper. QR codes are inexpensive to create and only need a camera-enabled smartphone to scan. QR codes provide pertinent and relevant information and deliver it quickly and efficiently.

Besides payment, QR codes also serve a channel of communication from store to customers. QR codes direct customers instantly to the website link, SMS, text messages of the physical store. Increasingly Chinese retailers have started using QR codes on billboards, posters, and flyers to offer discounts and product information. By scanning the codes, smartphone users can use mobile payment options to purchase the product or service immediately – promoting impulsive purchasing.

Once on the website, store owners can deliver any information they want, such as the product information, promotion, coupon, discount, product display, restaurant menu, price list, the location of the other franchise stores, membership, coming events etc. Customers can enter the virtue queue for train ticket or seats in a restaurant, or make purchases or orders from QR code. Receipts often contain QR codes, which open the portal of a wealth of information about the store or business, including ads, coupons, promotions etc. QR codes can allow easy access to events and consumers can download various calendars to their phones, ensuring they have the information needed to attend the event. Tipping digital content is quite popular in China, and content creators on WeChat's mobile publishing platform use QR code to collect tips with payment transactions processed by WeChat Payment.

Recognizing this growing trend, PBOC recently revealed plans to regulate QR based payment technologies [360] and has authorized the China Payment & Clearing Association to draft standards for mobile purchases linked to QR codes. Favorable regulation will likely support the development of virtual credit cards, providing further stimulus for Fintech firms focused on digital payments.

The popularity of QR code is for both offline and online. Any QR code on the WeChat messaging app can be decoded with the touch of a finger. In the US, Amazon begins to experiment with QR code-scanning mobile payments at its Amazon Go locations.

Besides the choice of technology, the failure of Apple Pay to crack the Chinese market has another reason. As a third-party payment service, both Alipay and TenPay are cross-platform services that are open to users on iOS and Android and any type of phone. Android mobile operating system has 75% market share in China as compared to 24% of Apple's iOS. This allows them to reach much wider market using the smartphone apps. Apple Pay only works with iPhone that automatically excludes the majority of the Chinese market who use Android phones.

There is a large presence of a new segment of digitally perceptive consumers – the Gen-Y and millennial– who account for 45% of consumption. They are driving the online retail market and leading the charge in China's mobile payments adoption, with 66% of post-1990s millennial shopping and 54% banking via their mobile devices. A rising number of young Chinese consumers end up accessing financial services for the first time through Fintech-developed platforms, rather than traditional banks.

Although third-party payment transaction still involves banks, it cuts traditional banks off from relationships with merchants and retail customers. The banks merely process the transactions for the third party without establishing a relationship with the payment parties. It deprives the banks the potential of

---

[360] "China's centralbank to standardize QR code payment", http://www.chinadaily.com.cn/a/201712/27/WS5a43be15a31008cf16da3d82.html

Other blockchain applications

other mainstay businesses, such as loans, deposits, and investments. The emergence of such new payment platform offers significant opportunities for Fintech companies to gain substantial scale but can have a potentially devastating effect on the banking status quo in China.

Fintech firms, such as online-only banks like MYbank and WeBank with streamlined lending processes and innovative credit rating assessments, have broadened financial access for a large segment of the population often ignored by the traditional banks in China.

The traditional banks are fighting back with their own Fintech transformation. For instance, ICBC, the world's largest bank by assets, has been adapting quickly to the Fintech revolution. ICBC embarked research of advanced technology and the cultivation of technical talents by establishing seven innovation labs in ICBC's head office for artificial intelligence, cloud computing, the blockchain, biometric identification, Big Data and internet finance and blockchain-based financial trading system. It is ramping up efforts within the payments space to capture customer data. It successfully launched an e-commerce platform, e-Buy mall, which has grown to become one of China's largest e-commerce platforms. ICBC relied on its capabilities as a bank to facilitate e-commerce, payments, and forex. Its quick payment tool, ICBC e-Payment, already had 60 million customers in September 2015. ICBC also launched an e-based finance product system that offers payments, financing and wealth management services, with the largest local online revolving loan extending US\$259 billion to more than 70,000 SMEs.

Other banks have also collaborated with Fintech firms to launch digital initiatives. For example, the Postal Savings Bank of China (PSBC), China's largest lender by branch network with 40,000 branches, is working with Ant Financial's MYbank and Tencent in internet and mobile finance. Through such collaboration, both parties benefit. Banks will be able to reach a new segment of customers and the online banking capabilities of the large e-commerce players. Fintech companies can establish brick and mortar branch office without physically building it, allowing them to venture into O2O business.

Chinese Government has an open policy to promote financial inclusion of China's 234 million unbanked people living in rural areas and in the poorest neighborhoods. The collaboration between the online-only bank and traditional bank fulfills the desire of government policymaker.

Before there was an online third-party payment system, there was also an offline third-party payment system. It works like a bank but without a branch office. Its physical presence is a kiosk or an ATM-like machine. Lakala, founded in 2005, is China's largest off-line financial service provider. [361] It boasts 60,000 self-service payment stations in China at convenient stores, supermarkets, shopping malls, community centers, hospitals etc. People can pay utility bills, buy train or airline tickets, buy movie or show tickets, book hotels, even buy wealth management products at Lakala kiosks. It also offers mobile payment platform and POS devices like those used in restaurants and cross-border payment services.

In a few short years, China's third-party payment system evolves from the mere online shopping payment system to an omnipresent payment system replacing cash entirely. The US payment ecosystem is also shifting toward mobile. However, such a shift requires a fundamental overhaul of the current credit/debit card infrastructure and unseats the existing benefactors. This will also depend on the incumbents, e.g. giant credit card companies, to come up with an innovative solution for the mobile payment yet without giving up their exclusive positions in the existing system. The introduction of Fintech may have a huge impact on the mobile payment system.

On June 21, 2010, the central bank of China issued the administrative rules governing payment services by non-financial institutions. These rules are the first set of regulatory measures China has adopted towards non-bank third-party payment processors. It will fundamentally affect China's third-party payment service, an important feature, and integral link of e-commerce, and possibly the e-commerce itself.

---

[361] "Company overview of Lakala Payment Co. Ltd.", https://www.bloomberg.com/research/stocks/private/snapshot.asp?privcapid=35020357

Other blockchain applications

Section 10.07   Chinese healthcare market

In the healthcare industry, millions of medical records scatter around in different hospitals, clinics, doctor offices or even insurance companies. Such data form a complete history of a patient and can be vital for diagnosis when needed. Today, these medical data are fragmented and difficult to share. Even the patient does not have access to most of these data. Such low efficiency and transparency is a great waste of social resources. Enhancing sharing is critical to the healthcare industry. However, there is also a privacy concern about these data. Safeguarding these data so that only authorized persons can have the access to it is vitally important. Therefore, the healthcare data make an ideal case for the blockchain application.

In October 2017 Hangzhou-Yunphant, a blockchain startup based in Hangzhou, entered a strategic partnership with Inspur to integrate blockchain with healthcare data. Inspur Group [362] is a leading provider of cloud computing and an advanced IT product and solution, one of the largest IT enterprises in China and the world third largest server provider. It serves more than 80 countries and regions around the world.

Once implemented, the blockchain will allow multiple organizations to access peer-to-peer networks without worrying about data security and integrity through encrypted data transmission. The platform synchronizes, consolidates and shares medical data created by various parties in real-time. The Yunphant Network is an enterprise-class consortium blockchain platform, called the YunphantChain. It is targeting to improve efficiency and transparency of the medical industry through blockchain technology. It can be quickly deployed for large-scale user application scenarios. The system supports operations such as authority management, monitoring, and maintenance, online deployment of Chaincode and status query.

Section 10.08   China digital currency policy

---

[362] http://en.inspur.com/inspur/2225886/index.html

China is already at the vanguard of mobile payment. Digital currency would be a natural step forward. Unlike India, China is taking a more cautious approach to the digital currency. China's digital currency aims to integrate into the existing banking system, with commercial banks operating digital wallets for the central bank's currency. The Chinese digital currency under development would only use a distributed ledger in a limited way, different from the well-known cryptocurrencies, such as Bitcoin. The issuing bank can verify the ownership of digital currency to realize peer-to-peer cash transactions. The goal is not to replace the paper currency, but to have both digital and paper currencies circulating side by side.

China conducted a limited test of its national digital currency in June 2017.[363] Such a test is in line with anywhere in the world, with the exception of India, which has a very aggressive digital currency policy. So far, the Chinese government has not yet issued an official timetable for the potential launch date. The test will probably be conducted in a limited zone rather than nationwide. The earliest tests will involve prototype transactions between the digital currency and some of China's commercial banks. Since there is no guarantee that the digital currency will prove to be successful for its intended purposes or without the unforeseen negative consequences, the test will provide valuable experience.

The added benefits of the digital currency to the currently prevalent mobile payment are not clear. However, as China tries to internationalize its currency. A digital form of its currency (RMB) can be circulating globally while its third-party e-payment system may not be adopted overseas. The envisioned benefits of digital currency are: It can give the government greater oversight of digital transactions. Blockchain transactions are easily traceable, allowing for an easier time finding and eliminating corruption. With digital currency ledgers, it is possible to analyze data and draw economic insights in real time. This would certainly help the

---

[363] "China central bank has begun cautiously testing a digital currency", Will Knight, https://www.technologyreview.com/s/608088/chinas-central-bank-has-begun-cautiously-testing-a-digital-currency/

Other blockchain applications

government in the development of its broad plans and strategies. It can also facilitate cross-border transactions, as well as the use of the RMB outside of China. It could lower the cost of financial transactions, and make financial services more efficient.

By testing a digital currency, China is seriously exploring the technical, logistical, and economic and operational challenges involved in deploying digital money, something that could ultimately have broad implications for its economy and for the financial system.

Section 10.09   Regulating TPP in China

The Chinese government did not regulate third-party payment system (TPP) until 2017. In August 2017, the Chinese government issued a regulatory notice that it has established a Network Alliance Platform to regulate the third party payments in China. All the third party payment systems were required to migrate their private platforms to the Network Alliance Platform by October 15, 2017.

Starting from June 30, 2018, the new platform is to clear all the third-party payments. The third-party payment companies, such as Alipay, will no longer need to interface directly with different banks. Their only interface will be the Network Alliance Platform, which will in turn interface with all the banks. This is not unlike the Automated Clearing House (ACH) operated by the Federal Reserve Banks and the private Electronic Payment Network in the US.[364]

The Network Alliance Platform is organized as a company. The two biggest players, Alipay and TenPay each have 9.6% of shares. Central bank supervises the platform. When this takes effect in 2018, the Chinese third-party system will be regulated just as the credit card payment system in the US. This new mechanism of third-party payment will have a huge impact on the digital payment market.

---

[364] "Overview of the US payments, clearing and settlement landscape", Federal Reserve of New York, https://www.newyorkfed.org/medialibrary/media/banking/international/03.Overview-US-PCS-landscape-Merle.pdf

There are winners and losers. The winners are small third-party payment companies. They no longer need to establish a relationship with each individual bank. Through the Network Alliance Platform, they have the same access to all the banks as the large players. The losers are the large third-party payment companies.

When the payment fund parks in the Network Alliance Platform, it does not generate interest for the third party. Even though the holding period is short, the amount can also be staggering. Such an interested earned used to belong to the third party, but now belongs to the NAP. In Q2, 2017, the China third-party mobile payment topped 27 trillion Yen, or $4 trillion dollars. If the fund stays in the NAP for three days average, the parked amount at NAP is $45 billion.

In addition, the proprietary Big Data, which used to belong exclusively to the third party payment companies, now have to be shared with the Network Alliance Platform.

Third party payment services have alleviated most of these problems and thus have contributed significantly to the growth of e-commerce in China. There are, however, problems associated with third-party payment services. A serious concern is that it could become a money-laundering tool. For example, a recent case in Suzhou involves a massive illegal gaming scheme operated by overseas online gaming companies. The perpetrators used third-party payment services to transmit US$ 520 million promoting the scheme. One of the masterminds behind the scheme, Keith Mei, was a senior manager in 99Bill, a leading third-party payment service provider in China. According to some reports, 99Bill may have reaped US$ 2.5 million in illegal profits from the scheme.

There is an urgent need for the legislation to address these problems. The legislation covers online payment services, the currency conversion service, issuance and servicing of pre-paid purchase cards, and other payment services conducted by nonfinancial institutions. Providers for these services must now apply for and obtain a "Payment Service License." The legislation also sets the threshold credentials for entities qualified to provide third party payment services and set credentials for entities controlling licensed third-party payment companies. Some estimate that about half of the 300 plus third-party payment

Other blockchain applications

companies currently operating in China could not meet these standards.

There are also new operational standards for third-party payment companies. These standards address various concerns associated with third-party payment services: security of customers' funds, money laundry concerns, and circumvention of the better-regulated banking system.

Chapter 11 A glimpse of the future

In the last five decades, it is Moore's law of semiconductor that drove the development of electronics industry. The doubling of the number of transistors on the chip of the same size every year made it possible that computers, communication equipment, and the internet have the power, as we know today. As the transistor is now ten times smaller than a virus, Moore's law comes to the end. The law of physics, which operates the transistor, breaks down quickly at this transistor size. In addition, the cost of building a manufacturing facility, wafer fab, is skyrocketing to tens of billions of dollars. It becomes both physically impossible and economically infeasible to continue the feast of shrinking transistors. Many are wondering whether the momentum of technological progress that we enjoyed in the last five decades will slow down or even end. Nevertheless, we look around, not only the technological progress is not slowing down, but also it is accelerating.

Instead of pushing vertically for the CPU speed, new applications are being developed every day to broaden the scope of applications. Such is the case of wide-spread use of internet and communication tools. Now the Big Data and artificial intelligence are opening up a new frontier of possibilities. The data-centric applications fit particularly well in one of the computational architectures – the Graphics Processing Unit or GPU. GPUs were originally designed for graphics applications, or gaming, in particular, to processing a large quantity of data fast. The emerging Big Data and artificial intelligence applications propel GPU toward the forefront of computational applications. GPU demonstrates faster speed than the traditional CPU.

A glimpse of the future

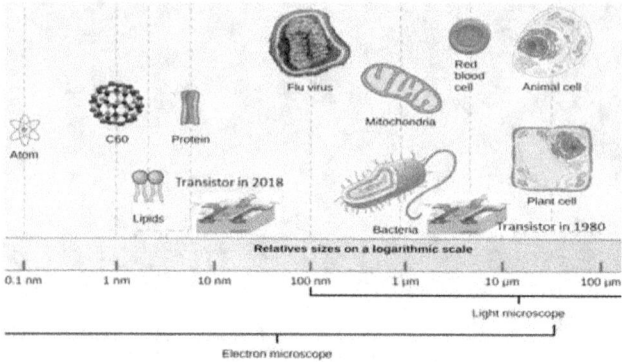

Figure 11-1 Dimension of transistor as compared to atoms, protein, virus, bacteria and animal cell

In the coming data-centric environment, the computing is shifting from instruction focused on data focused. One example is the Nvidia's CUDA parallel computing platform. [365] Intel, the world largest microprocessor manufacturer, is also pushing ahead their version of neuromorphic chips, which has the capability of self-learning. [366] It mimics brain's neural networks to relay information with pulses, to modulate the synaptic strengths and to store these changes locally. Further, Intel has a quantum computer chip in the making, which promises to boost computing speed by million folds.

Fintech, including blockchain, is just one of the many applications of the Big Data and artificial intelligence. Fintech will provide radically different innovation to the traditional financial industry that it will deeply affect the traditional financial services model. The direction and depth of the change may pose major

---

[365] "Moore's Law CPU scaling "is now dead" claims NVIDIA VP; GPU parallel computing is the future", Chris Davies, https://www.slashgear.com/moores-law-cpu-scaling-is-now-dead-claims-nvidia-vp-gpu-parallel-computing-is-the-future-3083858/

[366] "Intel's New Self-Learning Chip Promises to Accelerate Artificial Intelligence", Dr. Michael Mayberry, https://newsroom.intel.com/editorials/intels-new-self-learning-chip-promises-accelerate-artificial-intelligence/

challenges to the financial institutions today. It is a paradigm shift. With the baseline technologies accelerating at ever-faster speed, it bounds to have a drastic impact on the development of Fintech.

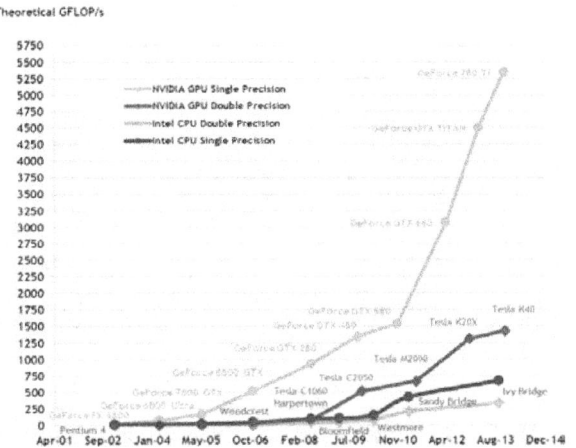

Figure 11-2 The progressing of computing power in GFLOP/s

However, no matter how Fintech develops, it cannot violate the basic principle: the maintenance of financial system stability and market order, the protection of the legitimate rights and interests of consumers. That means the regulation must keep up with the Fintech development.

Section 11.01   Pillars of the future technology evolution

Fintech is more than just an application of the blockchain technology. Many other technologies as well propel the Fintech development. Graphically, we can see how Fintech, blockchain and other basic technologies are relating to each other as shown in Figure 11-3. At bottom, there is the infrastructure layer. This layer is the foundation where all derivative technologies and applications are built on. This layer includes internet itself, which consists of P2P network and cloud. The IoT is the interface between the internet and physical world objects. Communication is also indispensable because it allows the transfer and sharing of all digital data. As the speed and bandwidth of the communication

A glimpse of the future

technology improve, more data transfer at a faster rate, and give birth of the Big Data application.

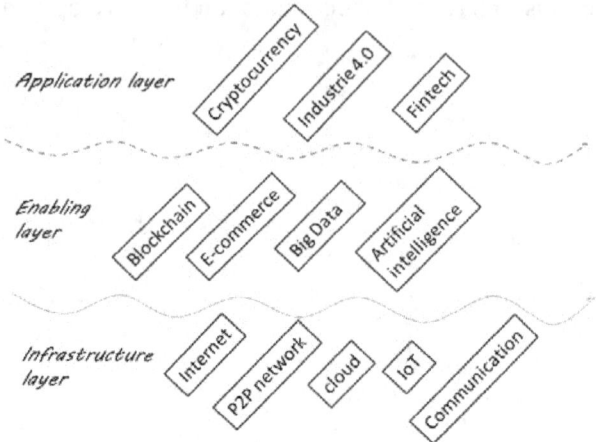

Figure 11-3 Elements of the future technologies

The middle layer is what we call the enabling layer, which enables all kinds of applications using the infrastructure technologies. They are the software building blocks, which reside in the foundation and from which all the applications are built. They are blockchain, all the e-commerce platforms, Big Data and artificial intelligence.

The six major components, Artificial intelligence, Blockchain, Cloud computing, Data, E-commerce, Fintech, are conveniently lumped together also known as ABCDEF.

On the very top are the application layers. The three major categories are cryptocurrency, Industrie 4.0 and Fintech. This book focuses the discussion on the blockchain and Fintech. However, the Industrie 4.0 revolution is also built on the same foundation.

Such a graph is conceptual and illustrative. It is one of the many ways to illustrate the relationship between many technologies. The technologies in the enabling layer are complementary to each other. The combination use of these technologies makes it possible to develop many applications in the application layer.

The graphic illustration is neither exclusive nor inclusive. For example, Industrie 4.0 requires additional infrastructures, such as sensors, robots etc, which are not shown here. On the other hand, one can argue that e-commerce is an application rather than baseline technology for the applications. Nevertheless, the graph serves a visual purpose of showing the important elements. In fact, the boundary between AI and Fintech is very blurry. The sweeping change in the landscape of data-driven business is taking place across all industries. The same AI technology, when applies to financial use, it is branded as Fintech. For example, when PayPal uses GPU-accelerated deep learning for fraud detection, it is a Fintech application.

Section 11.02   Data drive Fintech

The field of financial technology innovation is very broad. There are plenty of opportunities for asset management, loan business, insurance, payment, financial market operations and many others. Experience and establishment of the traditional financial institutions are also essential to the implementation of sound applications, such as risk management, supervision, and operation of the financial industry.

The traditional financial institutions took decades to build a stable and functional financial system, so will the new Fintech take. The eventual success of the Fintech depends only on when the new system can ride through several up and down economic cycles and navigates risks unscathed in credit, market, operation, legal, reputation, compliance, and liquidity. As the system evolved over time with accumulated experience and effective control, the market will get more confidence in the new system.

One of the major driving forces of the Fintech development is the Big Data. After all, all transactions, financial deals are data. Big Data technology includes infrastructure, modeling, collection, mining, and use of data.

Big Data, blockchain, and artificial intelligence will play an important role in the risk management beyond traditional methods. In the traditional financial system, as exemplified in the Chinese market as well as elsewhere, the SME and the general public low-income groups have a higher risk, except the credit

A glimpse of the future

card and mortgage business, despite their financial demand is strong. This is all going to change with the Fintech, which is well suited to tap this huge potential market using the mobile internet, Big Data, artificial intelligence and cloud computing.

Fintech empowered financial industry can provide solutions to the risk, credit intermediary, payment settlement and other issues, thus allowing them to reduce the cost of financial services, at the same time without taking additional risk. Fintech can bring the underserved and underground financial market into the mainstream financial market, thus greatly increases the market size. For example, the illegal gambling market in the US alone is US$150 billion annually. [367] Not to mention the product counterfeiting market is US$225 billion. The blockchain based supply chain management can effectively eliminate the product counterfeiting. This is one of the many ways that Fintech can generate significant economic returns. Not only the customers can benefit from lower interest, but also financial institutions can generate more return by reducing the cost of default, and serving a bigger market. The financial resources that underserved customers obtain can further fuel the economic growth.

Robot investment management is one of the many applications of the artificial intelligence. A low-cost intelligent investment advisor platform can manage the pension savings. Artificial intelligence based Fintech can direct savings into the most needed investment for the economic growth, R&D of critical technology, etc. instead of creating a real estate bubble, and providing a higher return to the savers.

Online robots consultants can tailor services for each customer to design a suitable investment plan based on the risk appetite, duration of the investment, objectives, etc. The service can cover the entire lifecycle of the customer, from infant to old age. The application of data technology makes financial services universal and customized.

Big Data management based on scenarios, customer behavior, and consumption pattern can explore new business models: capturing and analyzing data based on the customer-

---

[367] https://www.statisticbrain.com/black-market-illicit-trade-statistics/

related mass service information, mining and sharing big data can lead to other innovative financial services and cross-industry innovation, resulting in efficient and personalized service.

Consumption habits and consumer demand change over time. The consumption of financial services will have to adapt to the change. In the increasingly fierce competition in the financial environment, Fintech adaptors will be able to create a strategic differentiation to outwit the competition.

All financial transactions, including stocks, private equity, bonds, hedge funds, and derivatives can form immutable chains. The blockchain technology can find application effectively in the register, identification, and transfer of all kinds of assets and contracts.

Security is the prerequisite for the coming data powered economy. Network security laws and data security supervision will be the strategic level of national security. Network security will play a prominent role in the future economic construction, production, and operation of enterprises. It is safe to say that insufficient data security will compromise economic development in the data-driven economy. In early 2018, we have witnessed the impact of the Facebook data leak, which may have influenced the political outcome. [368]

Data and network security include a wide range of scopes: technology, management, awareness, business processes and many others. It also involves all aspects of an enterprise's operations, making it difficult to achieve full-scale security since all of the above-mentioned scopes are changing. The security work has brought and will bring new challenges as the technologies evolve and the scope widens. Mobile Internet, cloud computing, big data and other new technologies all have different security challenges.

Most of the financial services at a global level are exploring the benefits of data analytics and big data. In the next few years, technology will allow financial services to gather huge amounts of data to tailor experiences, services, and products to customer needs. Banks will be able to customize products and

---

[368] https://www.aljazeera.com/news/2018/03/cambridge-analytica-facebook-scandal-180327172353667.html

A glimpse of the future

services based on advanced customer profiling data. Big Data and cloud will help the industry move towards a better-connected world. Multi-organizational MDL's with smart contracts and IoT sensing will allow semi-intelligent, autonomous transactions. Ledgers with a smart contract can provide immutable records for the long-term and rely heavily on cryptographic techniques to function.

Big Data and Artificial Intelligence are closely linked. In fact, Artificial Intelligence is the application of the Big Data. Using Artificial Intelligence on Big Data can provide insights from advanced analytics without requiring traditional data scientist expertise. In this way, Big Data and Artificial Intelligence form a mutually enhancing loop. Such a technique is the smart data discovery. Smart data discovery will accelerate business intelligence capabilities and enable a new generation of data-driven decision-making. Using smart data discovery can lead to more innovative financial services.

Section 11.03   Communication driven Fintech

Mobile devices account for around 80% of all internet use globally in 2018. Telecommunication technology is in a race to meet the need network speed and bandwidth of the evolving Big Data, Fintech, IoT and AI applications. Fortunately, the 5G telecommunication technology deployment is right at the corner.

5G network and devices will have breakthrough improvement in speed, latency, bandwidth, power consumption over 4G. Using 5G, the smartphone can download at over 1GBps, and eventually rise exponentially to 20GBps. However, 5G's ultra-low latency is as important as its speed. Latency is the time it takes for a device to interact with the remote server. In a 4G network, the latency is around 50 ms, but in 5G, it is 1ms. In other words, a 5G device can respond 50 times faster than a 4G device. This will allow 5G providers to meet growing demand for data-intensive services, like streaming video.

In 2016, there were 23 billion devices on the 4G network. By 2020, when the deployment of the 5G network is complete, there will be 50 billion devices on the network according to the forecast. 5G will connect more devices at low power, at low cost, and with high reliability. No matter where you are in a city, the connection will always be on. This should lead to a surge in the applications, such as Fintech, which requires a network connection.

For example, China Merchants Bank (CMBC) is cooperating with Huawei to make use of its agile information and communication technologies to help strengthen its customer-centric experience. Using Big Data and cloud technologies, CMBC is able to integrate data in real time, cut the margin of error in financial assets by half, increase conversion rates and speed up personal credit reporting. As a result, customers are able to experience on demand and real-time services, no matter what platform they are using or location they reside in.

With the IoT growing, payment technology will surge on the back of phones, smart watches, earphones, activity bands, virtual/augmented reality headsets, flexible sensors on smart clothes, and smart glasses and eyewear, and many others. All these devices will have a real-time link to the cloud, and they will link to each other in their vicinity. This is the MIMO communication. MIMO is short for Multiple Inputs and Multiple Outputs. It is one of the capabilities of 5G, which will replace Bluetooth, RFID and many other short-range communications today. These devices will talk to each other, and swap data. That will enable payments from any wearable.

The critical data will be stored in MDL's, including identity, transaction, and content. The sharing of these data will create a powerful feature for Fintech. Identity authentication of the wearer will establish all the trusted needed for the transaction. 5G will also enable the micro-payment system, in which the unit of payment can be very small. For example, the parking fee unit can be as small as US$0.20 per minute, instead of US$12 per hour. Alternatively, the Wi-Fi charges can be based on 1 MB per unit.

5G will also revolutionize stock market transactions. Speedy buying and selling is everything in the stock exchange, and fractions of seconds can make huge differences worth millions of dollars a year. 5G's latency will settle the transaction instantly.

Fintech companies can make use of 5G's reliable connectivity and vastly improved download and upload speed, as

A glimpse of the future

well as the exponential cloud computing power to develop new products. For example, banks can develop personal banking assistants using artificial intelligence, cognitive computing, and machine learning to produce financial and wealth management products tailored to customer's needs. Such a robo-advisor service is already available today but will be more powerful and more customized in the 5G era.

The faster, more responsive, more pervasive wireless coverage of 5G networks will provide the backdrop for breakthrough changes in numerous industries.

The success of UK's Atom Bank, a bank based on the mobile app, illustrates a trend. The online banking is unstoppable. In parallel, many assets are moving online. According to some estimate, by the end of this century, the blockchain platforms may contain more than 50% of the assets in the world.[369]

Recently, VISA has unveiled an interesting concept of how mobile technology can change the way people shop. If someone finds something interesting to buy, he can simply take a picture of it and immediately receive the necessary information about this product for purchase. This is possible with the ability to overlay digital information on live images via a smartphone camera.

Section 11.04   Conclusion

The blockchain technology promises to revolutionize the financial technology. This is possible by the advancement of the Big Data, Artificial intelligence, communication technology and many other breakthrough technologies. Today, we are at the threshold of such a revolution. We can predict that in the next two decades, there will be an avalanche of innovations, which will change our day-to-day life dramatically. It will have a huge impact on the society, business, economy, even the balance of power.

On one hand, these new technologies will be an equalizer. It promises to bring the financial services, which previously cater the upper class, to the mass of the population. On the other hand,

---

[369] https://knect365.com/5g-virtualisation/article/a2ba2cab-6977-4858-a500-e4f8392dde51/5g-and-fintech

they will make the entry barrier to entering the business much higher. Only the biggest and the most well-entrenched entities will be able to carve out a piece of pie in providing the infrastructure, services, and business. Along the way, they will rip most of the benefits out of these technologies. The world will be divided between the economies with the technology and without the technology.

Likewise, the privacy and security will be another big polarization. On one hand, the blockchain technology promises anonymous transactions, private and secure. On the other hand, digitization of identity data, asset and much other personal information online, including health data and DNA will turn a person into a number.

Like many other technologies before our time, whether the coming disruptive technology revolution will bring greater benefits or disadvantage to the humanity is unknown. The unstoppable trend will surely continue and accelerate.

Since the pace of this technological revolution or evolution is moving extremely fast, the content of this book cannot keep pace with the change. Every day, there are new announcements and new development to the scene, this book can only serve as an introduction to whom are interested in the subject. The Internet itself provides a wealth of information on the subject of matter. With the information presented in this book, you will be able to follow most of the news on the subjects of the blockchain, Fintech, and many others easily and with interest.

Bitcoin, Blockchain & Fintech

Figures

Figure 1-1 Definition of centralization by Vitalik Buterin 19
Figure 1-2 Degree of centralization 22
Figure 1-3 A printed Bitcoin Public Key and Private Key 27
Figure 1-4 Year by year Bitcoin price 28
Figure 2-1 Bitcoin genesis block 37
Figure 2-2 A ledger 37
Figure 2-3 An endorsed check 38
Figure 2-4 Satoshi's illustration of Bitcoin transaction in this original publication[2] 40
Figure 2-5 Digital signature flow 41
Figure 2-6 Digital signature verification 42
Figure 2-7 Multisig for enhanced security 44
Figure 2-8 Hash and fingerprint 45
Figure 2-9 Examples of hashing using the demo program 47
Figure 2-10 Merkle tree 49
Figure 2-11 Block header 50
Figure 2-12 Difficulty of Bitcoin network 52
Figure 2-13 Bitcoin supply 55
Figure 2-14 Armory wallet 56
Figure 2-16 P2PHK payment flow 57
Figure 2-17 P2SH payment flow 57
Figure 2-18 Multiple inputs and outputs of a Bitcoin transaction 61
Figure 2-19 A brain wallet 64
Figure 2-20 A mobile wallet 65
Figure 2-21 Google authenticator 69
Figure 3-1 Number of Bitcoin transactions per day vs time 72
Figure 3-2 Bitcoin transaction confirmation time in minutes from 4/27/2016 to 1/15/2018 73
Figure 3-3 Distribution of transaction data size 74
Figure 3-4 Average transaction size 75

A glimpse of the future

Figure 3-5 Blockchain size in MB...81
Figure 3-6 Bitcoin transaction fees...83
Figure 3-7 Bitcoin transaction confirmation time in minutes.......84
Figure 3-8 In a softfork, new rules are subset of old rules. In a hardfork, the opposite is true. 88
Figure 3-9 The difference between softfork and hardfork............90
Figure 3-10 Bitcoin split..96
Figure 3-11 Payment from Alice to Bob through third parties...102
Figure 5-1 A Ripple network..127
Figure 6-1 Example of a shared MDL...159
Figure 6-2 Corda transaction consumes existing states and produces new state..175
Figure 6-3 A P2P internet..185
Figure 8-1 An IoT connected object..221
Figure 8-2 Tangle data structure (source: https://iota.readme.io/docs/what-is-iota) 224
Figure 11-1 Dimension of transistor as compared to atoms, protein, virus, bacteria and animal cell...340
Figure 11-2 The progressing of computing power in GFLOP/s.341
Figure 11-3 Elements of the future technologies........................342

Index

2FA 69, 70
Access 22, 159, 210
ACChain152, 153, 154, 201
Alibaba 136, 207, 208, 209, 261, 278, 279, 282, 283, 298, 299, 300, 302, 305, 306, 307, 314, 315, 318, 320, 321, 322, 323, 325, 326, 328, 329, 331
Amazon . 32, 199, 261, 277, 279, 282, 302, 333
AML 149, 189, 211, 212, 213, 265, 268, 271
Anti-Money Laundering*See* AML
APIs 149, 255, 262, 290
Application programming interfaces .... *See* API
Big Data5, 7, 150, 219, 221, 222, 236, 240, 252, 255, 260, 263, 273, 278, 283, 284, 285, 297, 298, 301, 302, 304, 307, 312, 313, 314, 316, 322, 323, 324, 334, 339, 341, 342, 344, 345, 346, 348, 349, 350
Bitcoin1, 21, 23, 24, 25, 26, 27, 28, 29, 30, 34, 37, 38, 39, 40, 41, 42, 44, 45, 46, 48, 49, 50, 52, 54, 55, 56, 57, 58, 59, 60, 61, 62, 63, 65, 66, 68, 73, 74, 75, 76, 77, 78, 79, 80, 81, 82, 83, 84, 85, 86, 87, 88, 89, 91, 92, 93, 94, 95, 96, 97, 98, 99, 100,☐101, 102, 103, 104, 105, 106, 107, 110, 111, 112, 114, 118, 123, 124, 125, 126, 127, 129, 130, 131, 132, 133, 134, 135, 136, 137, 138, 140, 146, 147, 148, 150, 151, 152, 155, 157, 158, 162, 164, 168, 173, 174, 175, 176, 179, 187, 188, 195, 209, 210, 211, 213, 219, 229, 241, 254, 256, 258, 295, 296, 337
Blockchain1, 25, 26, 31, 32, 33, 34, 48, 55, 77, 98, 99, 107, 133, 134, 150, 152, 154, 171, 189, 191, 193, 194, 196, 198, 199, 203, 206, 209, 212, 215, 216, 219, 224, 225, 229, 230, 234, 235, 238, 239, 240, 242, 243, 244, 245, 246, 247, 254, 255, 256, 257, 258, 260, 269, 280, 303, 320, 323, 324, 325, 337, 344
Centralization 20, 22
Cold Storage 65
cryptography 27, 29, 130, 132, 133, 189, 257, 267
Decentralization 20, 257
digital currency6, 15, 23, 28, 31, 77, 107, 117, 124, 146, 153, 158, 168, 169, 170, 171, 188, 191, 254, 258, 272, 281, 327, 336, 337, 338
DSS 44
ECDSA 44
Ethereum20, 23, 32, 33, 34, 93, 110, 115, 119, 123, 124, 133, 134, 135, 136, 137, 138, 139, 140, 141, 142, 143, 144,

A glimpse of the future

145, 146, 148, 149, 151, 152, 155, 157, 158, 162, 164, 172, 178, 181, 182, 183, 188, 204, 219, 224, 228, 232, 239, 240, 241, 242, 247, 248, 249, 254, 273, 303, 320
Fintech ... 1, 16, 24, 31, 224, 235, 236, 252, 253, 254, 255, 256, 258, 260, 261, 262, 265, 277, 278, 280, 284, 285, 286, 288, 289, 290, 291, 294, 295, 296, 297, 298, 300, 301, 302, 303, 305, 306, 307, 308, 309, 310, 312, 313, 314, 315, 316, 317, 318, 320, 321, 322, 323, 324, 328, 329, 330, 333, 334, 335, 342, 343, 344, 345, 346
Google . 26, 32, 70, 71, 253, 277, 279, 286, 289, 290, 302
hardware security module .... See HSM
Hash . 47, 48, 49, 58, 59, 70, 139
Hash based authentication code .... See HMAC
HMAC 70
HMAC-based One-time Password .... See HOTP
Hot Storage 65
HOTP 70
HSM 64
HydraChain 34, 178, 181
Hyperledger 21, 31, 34, 164, 182, 183, 184, 185, 187, 219, 230, 245, 259, 303
IBM 119, 150, 163, 171, 182, 183, 185, 189, 219, 223, 224, 243, 244, 245, 253, 288, 289, 305, 323, 325
ICO .. 32, 33, 152, 153, 207, 272, 273

Industrie 4.0 . 221, 225, 253, 304, 345
Initial Coin Offering See ICO
Intel 32, 182, 183, 184, 253, 287, 288, 289
Intellectual property 27
Internet of Things . See IOT
Internet of Value ... See IOV
IOT 221
IOTA Foundation . 225, 227
IOV 16
KYC 149, 169, 171, 189, 211, 212, 213, 257, 268, 269, 271, 315
Linux Foundation .. 31, 183, 245
MDLs ... 160, 161, 162, 163, 164, 165, 166, 171, 172, 173, 187, 189, 219, 236, 267, 268
MFA 69
Moore's law 341
MtGox 66, 80, 81
Multi-factor authentication .... See MFA
OpenBlockchain 183
P2PKH 57, 58, 59, 60
P2SH 45, 57, 59, 60, 93
Pay-To-Public-Key-Hash See P2PKH
Permissioned blockchain 20, 21, 22, 24, 27, 34, 79, 157, 158, 159, 172, 173, 181, 183, 199, 232, 256, 262
POW 50
private keys .. 28, 45, 57, 65, 67, 80, 88, 93, 99, 115, 130, 139, 234
Proof of Work .... See PoW
public key 28
R3 .... 21, 31, 175, 182, 197, 198, 259, 324

RSA 44
Satoshi Nakamoto 27, 28, 29, 30, 38, 41, 57, 78, 101
SegWit 45, 78, 89, 93, 94, 95, 96, 100, 101, 102, 104
SHA256 .... 48, 50, 124
Simple Payment Verification .... *See* SPV
SPV 62, 82, 94
T2S 275, 276
Time-based One-time Password .... *See* TOPO
TOTP 70
Two Factor Authentication .... *See* 2FA
Unspent Transaction Output *See* UTXO
UTXO 62, 151
Validation 22, 268
wallet 28, 38, 45, 58, 59, 65, 66, 67, 68, 69, 84, 93, 96, 99, 101, 116, 139, 147, 195, 321, 326, 327, 328, 331
Zero Knowledge Proof. *See* ZKP
ZKP 60, 61, 123, 126

www.ingramcontent.com/pod-product-compliance
Lightning Source LLC
Chambersburg PA
CBHW071041240526
45471CB00014B/99